Beyond Backpacker Tourism

© **Mixed Sources**
Product group from well-managed
forests, controlled sources and
recycled wood or fibre
www.fsc.org  Cert no. SA-COC-002112
© 1996 Forest Stewardship Council
FSC

**TOURISM AND CULTURAL CHANGE**
**Series Editors:** Professor Mike Robinson, *Centre for Tourism and Cultural Change, Leeds Metropolitan University, Leeds, UK and Dr Alison Phipps, University of Glasgow, Scotland, UK*

Understanding tourism's relationships with culture(s) and vice versa, is of ever-increasing significance in a globalising world. This series will critically examine the dynamic inter-relationships between tourism and culture(s). Theoretical explorations, research-informed analyses, and detailed historical reviews from a variety of disciplinary perspectives are invited to consider such relationships.

**Full details of all the books in this series and of all our other publications can be found on http://www.channelviewpublications.com, or by writing to Channel View Publications, St Nicholas House, 31-34 High Street, Bristol BS1 2AW, UK.**

**TOURISM AND CULTURAL CHANGE**
*Series Editors:* Professor Mike Robinson, *Centre for Tourism and Cultural Change, Leeds Metropolitan University, Leeds, UK* and Dr Alison Phipps, *University of Glasgow, Scotland, UK*

# Beyond Backpacker Tourism

## Mobilities and Experiences

Edited by
Kevin Hannam and Anya Diekmann

**CHANNEL VIEW PUBLICATIONS**
Bristol • Buffalo • Toronto

To all of our children, may they all be able to travel

Library
University of Texas
at San Antonio

**Library of Congress Cataloging in Publication Data**
A catalog record for this book is available from the Library of Congress.
Beyond Backpacker Tourism: Mobilities and Experiences/Edited by Kevin Hannam and Anya Diekmann.
Tourism and Cultural Change: 21
Includes bibliographical references.
1. Backpacking. 2. Backpacking–Social aspects. 3. Sports and tourism.
I. Hannam, Kevin. II. Diekmann, Anya. III. Title. IV. Series.
GV199.6.B49 2010
796.51–dc22 2009051736

**British Library Cataloguing in Publication Data**
A catalogue entry for this book is available from the British Library.

ISBN-13: 978-1-84541-131-2 (hbk)
ISBN-13: 978-1-84541-130-5 (pbk)

**Channel View Publications**
UK: St Nicholas House, 31-34 High Street, Bristol BS1 2AW, UK.
USA: UTP, 2250 Military Road, Tonawanda, NY 14150, USA.
Canada: UTP, 5201 Dufferin Street, North York, Ontario M3H 5T8, Canada.

The policy of Multilingual Matters/Channel View Publications is to use papers that are natural, renewable and recyclable products, made from wood grown in sustainable forests. In the manufacturing process of our books, and to further support our policy, preference is given to printers that have FSC and PEFC Chain of Custody certification. The FSC and/or PEFC logos will appear on those books where full certification has been granted to the printer concerned.

Typeset by Datapage International Ltd.
Printed and bound in Great Britain by Short Run Press Ltd.

# Contents

# Contributors

**Professor Kay Anderson**, Center for Cultural Research, University of Western Sydney, Parramatta Campus, Building EM, Sydney, NSW, Australia. Email k.anderson@uws.edu.au

**Dr Claudia Bell**, Faculty of Arts, University of Auckland, Human Sciences Building, 10 Symonds Street, Auckland, New Zealand. Email c.bell@auckland.nz.ac

**Dr Robyn Bushell**, Center for Cultural Research, University of Western Sydney, Parramatta Campus, Building EM, Sydney, NSW, Australia. Email r.bushell@uws.edu.au

**Gareth Butler**, University of Sunderland, Faculty of Business and Law, Priestman Building, Sunderland SR1 3PZ, UK. Email gareth.butler-1 @sunderland.ac.uk

**Dr Scott Cohen**, Department of Tourism, University of Otago, Level 4, Commerce Building, Cnr Union and Clyde Streets, Dunedin 9054, New Zealand. Email scohen@business.otago.ac.nz

**Dr Anya Diekmann**, Faculte des Sciences, Universite de Libre de Bruxelles, CP130/02, avenue F.D. Roosevelt 50, 1050 Bruxelles. Email adiekman@ulb.ac.be

**Dr Mark P. Hampton**, Kent Business School, University of Kent, Canterbury CT2 7PE, UK. Email m.hampton@kent.ac.uk

**Professor Kevin Hannam**, University of Sunderland, Faculty of Business and Law, Priestman Building, Sunderland SR1 3PZ, UK. Email kevin. hannam@sunderland.ac.uk

**Dr Jeff Jarvis**, Graduate Tourism Program, National Centre for Australian Studies, Monash University, Clayton, Victoria 3168, Australia. Email jeff.jarvis@arts.monash.edu.au

**James Johnson**, University of Sunderland, Faculty of Business and Law, Priestman Building, Sunderland SR1 3PZ, UK. Email james.johnson@sunderland.ac.uk

**Kath Laythorpe**, University of Sunderland, Faculty of Business and Law, Priestman Building, Sunderland SR1 3PZ, UK. Email kathleenmargaretlaythorpe@hotmail.com

**Linda Myers**, University of Sunderland, Faculty of Business and Law, Priestman Building, Sunderland SR1 3PZ, UK. Email linda.myers@sunderland.ac.uk

**Cody Paris**, School of Community Resources, Arizona State University, Tempe, Arizona 4020, USA. Email cody.paris@asu.edu

**Michael O'Regan**, School of Service Management, University of Brighton, 49 Darley Road, Eastbourne, East Sussex BN20 7UR, UK. Email M.J.O'Regan@brighton.ac.uk

**Dr Victoria Peel**, Tourism Research Unit, Monash University, Caulfield East, Victoria 3145, Australia. Email vicki.peel@arts.monash.edu.au

**Professor Christian Rogerson**, Environmental Studies, University of Witwatersrand, Private Bag, PO Wits. 2050, Johannesburg, South Africa. Email: Christian.Rogerson@wits.ac.za

**Peter Welk**, Kapellenring 25, 79238 Ehrenkirchen, Germany. Email petewelk@yahoo.com.au

# *Acknowledgements*

A large number of individuals and organisations were involved in the various activities that have made the publishing of this book possible. The continued development of the Backpacker Research Group (BRG) owes much to the support of the administration of the Association for Tourism and Leisure Education (ATLAS) who continue to provide much needed organisational support. The BRG meeting at Himachal Pradesh University, Shimla, was, of course, crucial to the development of this book and our thanks go particularly to Professor Bansal and his colleagues who helped to organise and support this endeavour. Specifically, we would like to acknowledge the help of James Johnson, a research assistant at the University of Sunderland, for his help in formatting the book. We would like to thank the authors for their perseverance with this project and, of course, the respective authors would also like to thank their colleagues, friends and families for their warmth and support in the writing of this book.

# Preface

This is the third volume of the on-going research programme on backpacking developed by the Backpacker Research Group (BRG) of the Association for Tourism and Leisure Education (ATLAS). The BRG aims to act as a platform for the discussion and debate between researchers of backpacker travel worldwide. It follows on from the success of the first volume *The Global Nomad* (Richards & Wilson, 2004) and the second volume *Backpacker Tourism* (Hannam & Ateljevic, 2007) and seeks to further shape this area of tourism research.

The idea for this particular book was initiated at a meeting at Himachal Pradesh University, Shimla, India, in March 2008, attended by members of the BRG, including researchers from the UK, Belgium, Germany, the Netherlands, Poland, Finland, South Africa, Australia, New Zealand, the USA and Malaysia. Drafts of some of the chapters contained in this volume were presented at this three-day meeting and these generated continued discussion and updates on the nature, meaning and significance of backpacker tourism. The meeting also enabled feedback to be given to the contributors of this volume and help to develop the concept of the book. Subsequently, other authors who could not attend the meeting also submitted chapters that were refereed and accepted. At the time of writing, new meetings of the BRG are being planned and further information about ATLAS BRG activities can be found on the internet (www.atlas-euro.org).

Kevin Hannam and Anya Diekmann
Sunderland and Brussels, July 2009

## Chapter 1

# From Backpacking to Flashpacking: Developments in Backpacker Tourism Research

KEVIN HANNAM and ANYA DIEKMANN

## Introduction

The present volume is the third in a series of books that have discussed research into the development of the backpacker tourism market over the last 10 years. From drifters to backpackers, travellers to flashpackers, this introductory chapter will examine the recent research into backpacker tourism. This book, however, is not an exhaustive account of backpacker tourism research, nor is it meant to be. However, we believe that this particular volume adds significantly to the academic literature on backpacker tourism and independent travel more generally by enhancing the theoretical, methodological and geographical research on the subject. As we shall see, some of the key issues are the changing profile of the backpacker market segment, the adoption of new means of travel, the use of new technologies, as well as the creation of new spaces or enclaves.

While it is not the aim of the present collection to provide a comprehensive literature review of the research to date on backpacker tourism (see Richards & Wilson, 2004), this introduction firstly seeks to outline some of the conceptual developments in backpacker tourism since *Backpacker Tourism* was published (Hannam & Ateljevic, 2007a). Secondly, this introduction provides a summary of the contributions in the present collection.

## Flashpackers, Backpackers and Travellers

One of the key developments in backpacker tourism, in recent years, has been in terms of the notion of the 'flashpacker'. The so called flashpacker has emerged as a new and key constituent of contemporary travel and exemplifies the changing demographics in western societies

1

where older age at marriage, older age having children, increased affluence and new technological developments, alongside increased holiday and leisure time have all come together.

The flashpacker has thus been variously defined as the older twenty to thirty-something backpacker, who travels with an expensive backpack or a trolley-type case, stays in a variety of accommodation depending on location, has greater disposable income, visits more 'off the beaten track' locations, carries a laptop, or at least a 'flashdrive' and a mobile phone, but who engages with the mainstream backpacker culture. Or more simply defined on Travelblogs.com (2009) as, backpacking 'with style' or even, backpacking with 'bucks and toys'. It is also seen as 'doable' with children in tow – with one flashpacker couple recently advertising the birth of their 'flashbaby' while on their travels (Flashpackingwife.com).

Indeed, Jarvis and Peel (this volume) cite *The Future Laboratory*, who in 2004 identified 'flashpackers' as older travellers on career breaks who 'can afford to splash out on some of life's luxuries when the going on the road gets tough' (The Future Laboratory, 2004: 13). More recently, this phenomenon has been explicitly highlighted by the backpacker industry, as a major hostel company advertises: 'Looking to treat yourself but, considering the current economic climate, afraid to splash out? With some of the luxury hostels listed you can pamper yourself without breaking the bank...' (Hostelworld.com, 2009).

By contrast, the work by Cohen (this volume) also reminds us of a different phenomenon – the lifestyle tourist who, like the earlier 'drifter', still spends the majority of his or her life indefinitely 'on the road' engaged in the backpacker culture. For both the flashbacker and the lifestyle traveller, however, it has to be recognised that new technologies have transformed the ways in which they travel and engage with their home-place and their social ties, as Paris (this volume) demonstrates. Nevertheless, it should also be noted that the backpacker tourism market is still dominated by many younger and less affluent tourists who spend most of their time in what have become mainstream and even institutionalised backpacker enclaves in 'traditional' destinations.

## Structure of the Volume

The present volume, thus, complements the aforementioned two previous books. It adds new theoretical dimensions in terms of a focus on mobilities and experiences and also broadens the scope of the research geographically. Indeed the present volume includes, alongside the traditional backpacker destinations of Australia, New Zealand and

South Africa, research in lesser known backpacker destinations, such as Norway, Tanzania and Mongolia. The book also engages present research of how backpacker tourism feeds into other forms of tourism, such as volunteer tourism (Chapter 9) and heritage tourism (Chapter 11).

Many of the chapters draw upon ethnographic field research methodologies, such as participant observation and in-depth interviews with travellers or a combination of these methods with documentary sources. However, some chapters use quantitative survey methods to identify specific sub-segments and behaviours of travellers (Chapters 3 and 4). While the first five chapters set the broader conceptual frame-work, the subsequent chapters deliver mainly empirical profiles with a particular focus on experiences in different countries and continents.

Following this introduction, in Chapter 2, Mark Hampton critically reflects upon his own experiences of being a backpacker himself, as well as researching other backpackers. In so doing, he charts the development of backpacker tourism from the 1980s up to the present with the appearance of 'flashpackers'. In the first section of his chapter, he explores his own experiences of two major trips through Asia as a backpacker and in the second part, he reflects on himself as an academic tourist – who actually gets paid to travel – through the prism of several research trips to South East Asia as a researcher doing fieldwork.

In Chapter 3, Jarvis and Peel focus explicitly upon the flashpacker phenomena, which they define in terms of older backpackers with higher levels of disposable income, travelling on a career break. In their study, they describe how 'upmarket backpacking' has developed through changing demographics and motivations. Furthermore, they seek to examine the ways in which the flashpacker market segment, with their particular travel behaviour and expenditure patterns, may present a niche opportunity for sustainable tourism development in the Fiji Islands. They conclude that policy makers within host destinations such as Fiji need to recognise the emergent diversity within the general backpacker market segment and to find ways of supporting the local industry in addressing the new demands associated with 'flashpacking'.

In Chapter 4, Cody Paris approaches the flashpacker phenomena from another perspective. He looks into how backpacker tourism has harnessed recent innovations in communication technology and, in particular, how online social communities have added a virtual compo-nent to the diverse mobilities of backpackers. While the physical mobilities of backpackers are still as important to the backpacking experience, new virtual moorings have developed that allow backpackers to be fully integrated in their multiple networks and also to maintain a sustained state of co-presence between their backpacker culture and their home culture. Furthermore, Paris argues that the backpacker ideals of independence, freedom and physical travel are all enhanced by the virtual mobility of backpackers. Based upon data from a self-administered online questionnaire, his chapter explores the various roles that communications technologies play before, during and after backpackers travel.

In Chapter 5, Scott Cohen develops an alternative argument to the flashpacker thesis in his examination of contemporary 'lifestyle travellers'. He argues that within the backpacker market, there exist a small proportion of individuals for which travel is not a cyclical break or transition to another life stage. Instead, for these individuals, leisure travel can serve as a way of life that they may pursue indefinitely. Drawing upon lifestyle theories, in this chapter he conceptualises these individuals as 'lifestyle travellers', a less pejorative term than 'drifter' (Cohen, 1972) or 'wanderer' (Vogt, 1976).

The next two chapters highlight the ways in which backpackers utilise important spaces and enclaves in order to create significant experiences. In Chapter 6, O'Regan conceptualises the importance of the backpacker hostel for the framing of backpacker 'performances'. His approach is based on the idea that hostels are historically, discursively, symbolically and materially part of the backpackers lived 'socio-spatial practices'. He examines hostels as a key symbol of backpacker travel itself, where individuals perform and interact, create narratives and build an identity. The 'backpacker hostel' has thus become central to the reproduction and development of backpacking, linking multiple spaces and times together.

James Johnson broadens this analysis in Chapter 7 to include the spaces of travel itself, building upon ideas of mobilities in backpacking research. He focuses on backpackers travelling 'off the beaten track' around central and eastern Europe. He investigates, in particular, the backpacker spaces on train journeys and explores the tactics of back-packers to create and maintain privacy on public trains. Through ethnographic 'thick description', Johnson also notes the backpackers' bodily responses to the feel, speed and conditions on board trains. He concludes that like the space of a backpacker enclave, travel space is also 'dynamic' and there are considerable overlaps in experiencing moments of travel and of dwelling. Using the train to sleep off a hangover or to meet new travel partners provide just two among many examples of how moments of moorings and mobilities inform one another.

Similarly, 'off the beaten track' Claudia Bell in Chapter 8 examines budget backpackers visiting Mongolia. Like Johnson, she also focuses upon embodied experiences and corporeal movement, including the physical sensuous responses that tourists must adapt to as they travel. Indeed, based on her own ethnographic travel experiences, travel writing as well as travel blogs, Bell highlights the limits of the backpackers' 'comfort zones' in a country like Mongolia, where only very limited tourism development has taken place.

The sexuality of backpackers has largely been ignored in the research until now. Hence, in Chapter 9, Linda Myers explores the motivations for, and personal development through, travel for lesbian backpackers in New Zealand. From her in-depth interviews, two themes emerged from these women's travel experiences in New Zealand, namely, the need to escape the heterosexual world and its social constraints, on the one hand, and the importance of distinct lesbian spaces, on the other hand. She argues that New Zealand was chosen as a destination by the interviewees because it has specific provision for gay women in the form of accommodation and tours, it was a location that offered freedom, peace and safety and it was perceived as being gay friendly.

In Chapter 10, Kath Laythorpe examines how backpackers engage with volunteer tourism – a growing phenomena in the 21st century. This research analyses the motivations, expectations and experiences of considerable numbers of backpacker-volunteers travelling in Tanzania. Through interviews and participant observation, Laythorpe reflects on the advantages the backpacker-volunteers gain from their experiences. She concludes that the opportunity to combine both backpacking and voluntary work in these destinations is becoming increasingly common. Volunteer tourism offers the younger backpacker an institutionalised experience that can be contrasted with the more flexible and less constrained backpacking travel. The volunteer accommodation often provides a suitable enclavic space from which to explore and experience the culture of the area in a way that is often more intense than that of the average backpacker, but with the added safety net of a travel organisation providing reassurance and security.

Chapter 11 by Gareth Butler on Norway is the last chapter that looks at backpackers' experiences. Norway is perhaps an unusual setting in which to study backpackers, however, because of the high costs of travel in Norway, many tourists utilise the backpacker infrastructure such as hostels. Nevertheless, Butler analyses backpackers in Norway in terms of their personal heritage ties, their experiences of the vast, wild Norwegian landscape and their use of Norway as a 'platform' for exploring the rest of Scandinavia.

The next two chapters take a different perspective on the backpacker travel industry by focusing upon the supply management side rather than on the experiences of the backpackers per se. In Chapter 12, Peter Welk argues that the mainstreaming of backpackers and the mainstreaming of backpacker businesses go hand in hand. For a long time, the management of businesses in the backpacker industry was dominated by lifestyle entrepreneurs, who had been backpacking themselves in the past and saw an opportunity to combine their passion for the traveller scene with making a living. More recently, Welk argues, this pattern has been changing rapidly with many managers of backpacker businesses now 'dropping down' from the corporate world and introducing branding into the previously somewhat anarchistic backpacker business community. He shows how backpacker destinations can be created by corporates and how the original backpacker culture may be replaced by purposeful planning, blurring the distinctions between conventional tourism and backpacker travel. Welk argues that such backpacker destinations are increasingly commercially promoted, as is illustrated in his case study of the Town of 1770 in Australia.

Chapter 13 by Robyn Bushell and Kay Anderson looks again at the backpacker/flashpacker tourism industry in Sydney, Australia, focusing on the increasingly complex definitional and management challenges. Following the authors of the other chapters in this book, they highlight the changing diverse profile of backpackers. They examine Australia's policies of providing travellers with a temporary work visa, who then may reside in one place for several months. As a consequence, residential and backpacker tourist experiences become entangled, leading to some significant cultural clashes as the former regard the latter with some disdain.

Chris Rogerson closes the volume in Chapter 14 with an analysis of the strategic planning of backpacker tourism for an emerging destination in the context of the global backpacker tourism market. Only recently has South Africa recognised the strategic significance to promote this kind of budget tourism in the country. Rogerson highlights the development of backpacker tourism policies, the growth and structure as well as organisation of the emerging backpacker industry, including the appearance of planning initiatives at national and sub-national levels of government through surveys with backpackers and interviews with key stakeholders.

## Conclusion

Theoretically, many of the chapters in this volume have engaged with Hannam and Atelvejic's (2007b) call for an engagement with the new mobilities paradigm to provide analyses of backpacker tourist's experiences. With a broad range of case studies and geographical distribution, many chapters demonstrate comparable developments in the backpacker tourism market segment in recent years. With the advent of the contemporary flashpacker, the backpacker industry has continued to evolve both commercially and institutionally. Yet, as some chapters show, there remains a tendency for some backpackers to go beyond and to seek remote places for self-discovery and physical challenges. Moreover, these chapters also demonstrate the ways in which backpackers continue to engage with other forms of tourism.

Chapter 2

# Not Such a Rough or Lonely Planet? Backpacker Tourism: An Academic Journey

MARK P. HAMPTON

## Introduction

Much of the literature on backpackers is written from the perspective of the outsider, the researcher, the university-based academic working in the field who immerse themselves in the world of the backpacker. Some work is broadly positivist, but there is an increasing interest in using ethnographic-type approaches to understanding backpacker tourism. Some academic writers have noted, in passing, their own experiences as backpackers, but as with much tourism research, this tends to be 'one-shot' type fieldwork, case study-based, and although interesting and gives useful insights, has some limitations. This chapter is an attempt to interweave my own personal narrative as a former backpacker with the way in which the phenomenon, and its study, has continued to evolve over recent years; thus it takes the form of an explicitly reflective academic journey. In this, it has some similarities to autoethnography, where the writer metaphorically steps into the text. However, I am in good company as Hutnyk (2008) noted recently in his blog:

> The diary, a memoir, notebooks, letters from the field – the ephemeral residue of the research process of anthropology has increasingly attracted attention, become raw data for cultivation, sifting the soil.

This chapter is structured in four parts in a broad movement over time covering the last 20 or so years in approximately chronological order. The first part explores my own experiences of two major trips to Asia as a backpacker. The second part discusses the academic tourist through the prism of several research trips to South-East Asia as a researcher doing fieldwork. The third section explores what might be called way markers on the journey, that is, I discuss some emerging issues as backpacker tourism (and perhaps its nature) changes over time; and the final part concludes the chapter with some further reflections on the possible journey ahead, and makes some suggestions for further research. Travelling, journeys, mobilities, personal stories and narrative, both grand and immediate: these are the metaphors that run through this chapter.

## With the 'Freaks' in India

It is March 1984. With an entire year free before going to university, my 19-year-old self was setting off to travel alone around India and Nepal on a two-month trip. I set off with an old fashioned, steel-framed and rather heavy, green rucksack. At that time, the cheapest flight from London to New Delhi (apart from Aeroflot and I had heard some pretty hair-raising stories about them) was with Ariana Afghan Airways and cost £290 (around $600). In comparison, British Airways were then selling direct flights to India for around £500. I bought the tickets with some trepidation from a 'bucket shop', which is a deep discount travel agency based in London, but they arrived in good time in a thin white envelope. This was before the terms 'gap year' or 'backpacker' were at all commonplace. In fact, within India, the independent travellers were often nicknamed 'freaks' by the locals, especially the aging hippies, still then very much in evidence on the beaches of Goa and around the guest houses of Kathmandu (Iyer, 1988). They were, in a sense, the forerunners of the modern backpackers, and broadly conformed to what Cohen (1973) called the 'drifters', the remnants of the overland travellers of the late 1960s and 1970s who had pioneered (extremely) low-cost travel to, and around, Asia. Their presence had helped develop new destinations, such as Goa, Kathmandu and Kuta beach, Bali, allowing local entrepreneurs to start opening small-scale accommodation and eating places to meet this growing demand.

The flight to India was an indirect route and took over 20 hours. The first leg was in an old Ilyushin Il-62 aircraft belonging to Czechoslovakian Airlines (a former troop carrier by the look of it) to Prague, where we boarded the somewhat battered-looking DC-10 of Ariana Afghan, then on to snowy Moscow, then Tashkent, Kabul (under Russian occupation at the time with heavily armed soldiers with Kalashnikovs, MiG fighters and bulky Antonov troop transport planes), finally arriving in the dust, heat and smells of Delhi. Once there, I mostly travelled around by rail, including a wonderful steam train journey through Rajasthan, then by coastal steamer from Bombay (now Mumbai) to Goa, and finally by bus, winding up through the Himalayas to Kathmandu.

Even though I'd read up about India beforehand, at that time *Lonely Planet* was little known and I had not come across their seminal *India Survival Kit*. In fact, it was another traveller who highly recommended it, and so I bought my own Indian-printed copy that had pages censored with text blacked out that referred to sensitive issues such as the conflict with Pakistan and the Kashmir problem.

Like many young travellers, I religiously kept a diary of my trip, in this case written in a couple of cheap exercise books bought from my local Woolworths department store. At present, although you still see backpackers with bent heads concentrating on writing their stories in cheap diaries in traveller cafes in Ho Chi Minh city or elsewhere, you are as likely to see them rapidly typing on a laptop with a wi-fi link to a social networking site, or texting their friends or families back home with the latest news of their adventures. However, below I narrate my own backpacker travel 'stories' or 'reflective ethnography' from these early diaries.

Travel stories: I
We waited about an hour for the bus & when it finally arrived – wow. It was a stretched VW type minibus - & crammed in it were at least 25 people! We were squeezed into the boot section & spent the ½ hour ride bent double over baskets of fish, banging our heads on the roof at every bump! What a journey! (India Diary 3, 1984)

## Backpacking in Asia: 'I'm not a Tourist, I'm a Traveller'

We now fast-forward five years. It is now 1989 and next in the chronology after my first degree at university and some time spent working in various jobs, I took the opportunity to return to Asia before my PhD started. This trip through South-East Asia was for over four months, spanning the end of 1989 and the start of 1990. This time I travelled with my school friend Paul. Once again, air tickets were booked through a London 'bucket shop', flying to Asia with another airline, this time it was *Biman Bangladesh* for the trip London-Bangkok via Dhaka. Again, the journey from Europe was slow and indirect. However, this time, due to a change of airline timetable, we needed to spend two days in Dhaka before the final leg to Bangkok. *Biman Bangladesh* contacted us before we left England and asked if we wanted to change our flights given the unexpected stopover in Dhaka. If not, they would put us up in a modest hotel in the city. However, we were only too happy to have a fully-paid stay in the Bangladeshi capital en-route to Thailand.

On the flight in another elderly DC-10, amid the majority of Bangladeshi passengers, there were about a dozen Europeans and around half of us were backpackers heading for Thailand, comprising people mostly in our 20s and early 30s and a roughly equal mix of men and women. However, the other Europeans turned out to be a very different type of tourist indeed. They were all men, late middle-aged, and it became clear that they were in fact sex tourists (Opperman, 1999; Ryan & Hall, 2001; White, 2008). Some were dressed in tracksuits, and one even had a shirt open to the chest showing a gold medallion, and they were all heading for the infamous red light areas of Bangkok and Pattaya using the cheapest airline possible. Unfortunately for them, they had not been contacted by the airline about the extra two days enforced Dhaka stopover albeit with a free hotel and meals included. Their reaction was interesting to say the least. This group of British men just sat in the hotel smoking cigarettes and drinking glasses of cold Seven-Up and Coke and bemoaning the fact that Bangladesh was a Muslim country and that alcohol was impossible to get at this hotel. For the rest of us, we could not wait to dump our rucksacks at the hotel and get out exploring the city, first by pedal rickshaw, and later by chartering a small skiff on the river.

Once in South-East Asia, we travelled overland through southern Thailand, Penang (Malaysia) down through Sumatra, Singapore, back via Malaysia and then on to northern Thailand (Sukathoi, Chang Mai and the North-East river border with Laos). Later, I would think about this route and discuss it with my colleague, Amran Hamzah, as comprising part of the 'classic' early 1990s backpacker route (Hampton & Hamzah, 2008). Unlike my India trip five years earlier, this time I had bought the *Lonely Planet* guidebook – *South-East Asia on a Shoestring* – the so-called 'yellow bible' for backpackers. *Lonely Planet* by then was becoming very well-known as the alternative and trendy travel guide that we backpackers needed to have with us. As before, I kept a travel diary in several exercise books. These two excerpts give a flavour:

Travel Stories II
...as I've been trying to scribble these scattered thoughts, my attention keeps being distracted by the inevitable "hello meester" or "where you from?" Here, as in many cafes, urchins come up & pester you about a shoeshine. Difficult with sandals! (SE Asia '89–90, Diary 2)

Travel Stories III
Travel's fun, there's no question, but I hope it doesn't get too serious! If, as I wonder, there's a sub-culture developing, I hope the better

qualities like helping others out with info. etc & the good things of the travel mind-set stay uppermost. Otherwise it'll be no different to the tourists (YUK), we all despise & laugh at. "Same, same but different" – maybe? (SE Asia '89–90, Diary 2)

## 'You get Paid to Go There?' Academic Tourism

By the mid-1990s I was established in my first lecturing post at a British university and I found that my own interest in backpackers was itself changing. Although I still enjoyed visiting South-East Asia as a backpacker myself whenever I could save up the money to return, I found that I had a growing professional interest in starting to think about what impact backpackers might be having in less developed countries (LDCs). As I began what was to become later formalised in my academic writings as a 'literature review', I was struck by how little academic research existed at that point. What little literature I could find made a rather a short list comprising the seminal work of Cohen (1973, 1982a), Loker-Murphy and Pearce (1995), Pearce (1990), Riley (1988) and Rodenburg (1980). There was clearly much work still to do in understanding the backpacker phenomenon. My interest grew in attempting to discover more about backpackers in the field, and so I applied for some funding from my university to pilot backpacker research. This turned out to be the first of what would eventually become a series of journeys to South-East Asia as an academic tourist researching backpacker tourism.

The funding covered an initial research visit in 1996 to some small well-known backpacker islands off the coast of Lombok in eastern Indonesia. I had decided to focus on one particular island, Gili Trawangan, which had become a 'rest place' for backpackers in the emerging Indonesia route of the 1990s that broadly ran West to East (Sumatra; Java – with visits to Jakarta, Bogor, Yogyakarta, Mount Bromo; Bali; Lombok; Komodo and sometimes Flores). I was intrigued by the idea of a 'rest place', which is on the extended trip that characterises backpacker travel, of there being destinations that act as holiday sites within the longer holiday and travelling period. Specifically, the fundamental question was: were backpackers of overall economic benefit for the local host community in LDCs? In this, my approach was *not exactly* an anthropological approach, nor purely a market-led approach (Wilson & Richards, 2008).

At the time of doing the fieldwork, the Gili islands only had backpacker accommodation. Now there are boutique hotels and further construction is ongoing at the time of writing (August 2008).

One of the initial challenges I faced was the lack of existing data on tourism in the islands, let alone the backpacker sub-sector. This had two aspects. First, where official data on tourism did exist within Indonesia, it was almost impossible to use for my purposes, since tourism data for the Gili islands could not be disaggregated from data collected on the larger nearby island of Lombok, let alone separated from data concerning the whole province that it is located in – West Nusa Tenggara. Secondly, I soon found that even where data might have been useful, the preponderance of under-reporting of rooms, beds etc., due to fears of tax revenue collectors meant that data would be insufficiently reliable to use. Therefore, it became clear that I would need to spend time in the field collecting baseline data myself. I did this using a mixture of semi-structured interviews (backpackers and service providers), question-naires, participant observation and site mapping. The latter was either on foot in the heat or by bicycle. For example, no data existed on even the number of beds on the island. I simply went round counting accom-modation units and recording the number. Based on a reasonable assumption of double occupancy, I then double-checked my overall estimate of the number of beds with several key respondents who were able to help me triangulate my own estimate.

The main findings were not too surprising given my own prior experience in the region, namely that the backpackers were indeed making a key contribution to the local economy. What did surprise me when analysing the interviews ($n = 33$) and the small pilot questionnaire ($n = 47$) was the amount that they were spending. The backpacker expenditure was relatively high, around US$65–100 per week on food and accommodation. In the local context, this was higher than I had imagined and was broadly equivalent to an average Indonesian's monthly salary at that time. However, the study also raised some difficult questions of the increasing role and power of outsiders as resorts progress up the resort cycle. (These findings appeared as Hampton, 1998; and Hampton and Hampton, 2009.)

Sitting on the outside deck of the slow ferry from Bali to Lombok, we sat opposite a European backpacker who was reading *The Beach* (Garland, 1996). I was struck at the time by the ironies of seeing a backpacker on her own journey, reading a novel about imaginary back-packers on their journeys, observed by me on mine. Now, looking back and reflecting some more, there are further layers I can now question, was I then, at that point still a backpacker myself, a participant too as well as an observer, or was I just another academic researcher? There is a layering here, a mix of narratives, realities, mobilities even. Here, we start to touch upon the notion of the academic as participant in the story, effectively inside the phenomenon they are studying. This has a long history, stretching back to the founders of anthropology such as Evans-Pritchard in the Sudan and Malinowski in the Pacific in the early 20th century (Moore, 2004).

Within the emerging field of backpacker research, this could be explored more deeply, especially given how the presence of the researcher themselves may or may not affect the 'subjects', that is the backpackers. In my own narrative, at that time, I was a European academic who was then not much older than most backpackers that I wanted to approach. The majority of backpackers at that destination and my research site were mainly European or Australasian and, typically as later studies also confirmed, were usually highly educated to university or college level. As such, when I approached backpacker respondents to ask them to participate in interviews or questionnaires, the reaction was normally very positive. In fact, sometimes I ended up having long conversations after the interview or questionnaire was completed about the research, about how I ended up doing it, my own experiences as a backpacker etc.

Incidentally, as I write this passage in mid 2008, sitting in yet another hotel, this time in Malaysia, with the wisdom of hindsight it strikes me that this was a missed opportunity to have metaphorically 'kept the tape rolling', that is, to have kept on recording the conversation as additional material. However, in a sense I was lucky to be able to exploit who I happened to be as a researcher and my own travelling experiences, since obviously I had little choice over my own ethnicity, age or gender. It is interesting to consider whether another researcher of different age, ethnic origin, and perhaps gender to some extent, might not have got past the gatekeepers.

## Researching Backpackers: Yogyakarta, Indonesia

In 1997, I managed to access some more small-scale funding from the UK government's higher education funding council (HEFCE) to cover the costs of further field research on backpackers in South-East Asia. For this visit, I decided to work on an urban backpacker destination, the area of Sosrowijayan, an established backpacker enclave in the Javanese city of Yogyakarta. Unlike the research on Gili Trawangan, by necessity this fieldwork had different timings mainly due to the need for cover for my university teaching, so I needed to do the fieldwork in November, which happened to be low season, with less European backpackers around. This was not necessarily a disadvantage for doing fieldwork, since although there would be less backpackers around, the accommodation and restaurant owners and other staff would generally have more time to give in-depth interviews to a researcher. In addition, the wider context had also changed since the Lombok fieldwork only 18 months previously. Specifically, international tourism to Indonesia was significantly beginning to slow, reflecting the growing Asian economic crisis with worries about Indonesia's political stability and the poor air quality from the forest fire haze.

During these few weeks' fieldwork, I stayed in a small hotel on the edge of the backpacker enclave in Yogyakarta, which was jointly owned by a local and a European – but I recall thinking, was I still a backpacker myself? I did indeed travel with a rucksack, albeit a more modern internal frame-type by then, but rather than using a lower cost airline, I flew from the UK to Asia with a large scheduled airline this time and then took an internal flight to Yogyakarta rather than going overland from Jakarta. Another difference was that, unlike the earlier field research that mostly worked with backpackers themselves, I wanted to strengthen my understanding of the local impacts by interviewing the owners in more depth; therefore I would need some help. Colleagues at one of the established local universities helped me find a research assistant who could also translate the respondents' more complex replies. That too was different. In the earlier fieldwork on the islands, I probably mostly stayed in or close to the imaginary backpacker tourist bubble. Now in Yogyakarta, I was working more with Indonesian academic colleagues and my local research assistant. In retrospect, I was moving away from being with the backpackers, that is being *a* backpacker, to being more a researcher *of* backpackers.

Regarding results, this time the main findings were that in some fairly fundamental ways, hosting backpackers had transformed a previously poor *kampung* into a charming, well-cared for enclave of narrow lanes and small accommodation and eating establishments. The area had been described as reminiscent of old Kyoto in Japan. To my surprise, at times I found it an emotional experience and was quite moved hearing the stories, the emerging narratives, of how hosting backpackers had radically transformed some local families' livelihoods. One woman respondent told me with great pride in her voice and manner about how she could now afford to send her children to school and how the *kampung* had changed so much from being a poor district.

The other point that really struck me when doing the fieldwork was that the local tourism officials were clearly disinterested in backpackers. Over the course of the interviews, it became very evident that they were busy planning for the 'modern' redevelopment of the old city and urban *kampung* areas, and that the existing backpacker tourism and small-scale accommodation had no place in this vision of a large-scale redevelopment and the construction of new urban tourism infrastructure. Once back in the UK, the fieldwork was analysed and then written up (Hampton, 2003).

## Researching Backpackers: Malaysia

The chronology now moves to May 2006. Amran Hamzah, now a professor of tourism planning, won a major £50,000 project from the Ministry of Tourism Malaysia to examine the various impacts of backpacker tourism and consider its further potential. The commissioning of this large-scale study was a significant moment in the story of government interest in backpackers, marking the first major interest shown by a LDC in the Asia region (the only other government that I am aware of is South Africa, see Rogerson, 2007). I had helped write the proposal and so became a part of the research team.

Fieldwork took place over summer 2006 in Malaysia and the team visited all the main backpacker centres, plus there were brief research trips to the enclaves in Thailand (Kao San Rd) and Vietnam (Ho Chi Minh city). This project was also different as I was not working on my own with just one research assistant, but this time I was part of an international research team comprising three academics and two local research assistants. In addition, my family also came on the fieldwork, including my son who was then around four years old. It was interesting to observe that the backpackers, mainly Europeans, ignored him, whereas the local respondents and local team members positively loved him being around, which further enabled the research.

Findings from this major project have already appeared in the public domain regarding the economic impact of backpackers and planning policy implications (Ministry of Tourism, 2007); the spatial flows and changing geographies of backpacker routes (Hampton & Hamzah, 2008); and the political backpacker tourism in small islands (Hamzah & Hampton, 2007).

My most recent work is for a major two-year project funded by the British Council on the developmental impacts of scuba dive tourism in coastal Malaysia. Backpackers still appear in this new story, but play more of an incidental role. One research site selected was in the Perhentian islands off the east coast of peninsula Malaysia, a well-known backpacker destination since the early 1990s (Hamzah, 1995). It is also a major location where many learn to scuba dive, illustrating how backpacker expenditure is 'lumpy', or highly uneven over their trip with sudden high expenditure. Typically backpackers may live daily on a very low budget and try to bargain for most things to save a few dollars here and there, and then spend a large amount on a special experience such as learning to scuba dive in the southern islands of Thailand, or going hot air ballooning in Turkey (Tucker, 2003).

## Way Markers? Milestones in the Continuing Backpacker Journey

It has become clear that backpacker tourism has been changing over time, since it first emerged from its 'hippy' roots during the 1980s. In this section, I will briefly consider some of the new issues, or milestones, in its journey: massification; the diffusion of taste; and fragmentation.

Most recently, there seems to be a massification process going on in some regions such as South-East Asia, that is, an increasing number of European and Australasian young people who are seen travelling with backpacks. However, I would argue that this alone does not necessarily make them backpackers, and suggest that there may be now be a blurring of boundaries between types of youth tourists seen in such regions. Specifically, from my observations in my most recent research visits in 2006 and 2008, I am beginning to wonder whether the mainstream self-styled 'backpackers' are really any different now from the majority of other youth tourists?

The idea of a backpacker sub-culture expressed through certain markers, such as dress (wearing local ethnic clothes), soft (bounded) adventures and relative hardship (long tiring overland journeys on crowded local transport etc.) and the construction of a backpacker identity seem to be changing. Groups of young tourists with backpacks can now be observed wearing internationally fashionable clothing, travelling by private mini-bus from backpacker enclave to enclave, and sitting in large groups watching the latest film being shown in a backpacker place in the form of a pirated DVD. In part, some of this is supplier-driven as local entrepreneurs from the early 1990s created a parallel infrastructure to meet the demand, filling up mini-vans entirely with young westerners to drive them from Kuta beach up to Ubud, or from Penang to the Cameron Highlands etc.

But, does it matter if there is a massification process going on? At one level it probably does not necessarily matter that much in terms of the tourists themselves. For the local service providers, since they are so close to their market, they will have adapted to the changes already. In a more conceptual way, however, it could be that there may be a remnant of hard-core backpackers who appear to inhabit a binary construction versus what could now be called mass backpackers. Spatially, the new areas being opened up by the pioneers, such as parts of Vietnam and Cambodia, lend themselves to analysis as a constantly moving frontier, or moving edge, as more inquisitive or adventurous individuals are always looking, searching for new places, the process that Garland (1996) skewered in his novel as the search for the 'perfect' beach.

In addition to the massification process noted above, I would also suggest that at the same time there are also signs of the increasing fragmentation of the backpacker market. This now includes the 'backpacker plus' (Cochrane, 2005). This is an interesting development and this segment contains tourists who still may travel with a backpack, but who tend to have shorter overall trip durations (that is weeks, rather than months) compared with backpackers as commonly defined. They are often professionals travelling in their annual holidays or sabbaticals, and who have less need to survive on very low budgets and so prefer to stay in slightly more expensive accommodation. The accommodation providers have reacted to this emerging market with small, slightly more up-market guest houses springing up recently in parts of Kuala Lumpur and Penang for example.

Another trend that can be observed of fragmentation is the rise of national or ethnic groups of backpackers that may travel together or meet up using new technology at certain sites or accommodation. In particular, the rapid rise of Israeli backpackers in both India and Thailand has been discussed by Hottola (2005). Associated with this is the relatively new emergence of Asian backpackers, who are gradually appearing in the main enclaves and destinations in South-East Asia in particular. As yet, Asian backpackers are still the minority in other backpacker regions, such as Latin America, but I would expect that may change in the next few years if certain Asia-Pacific countries continue their rapid economic development.

## Conclusions

Backpackers have been studied by academics broadly since the 1980s with a boom in publications from the late 1990s onwards. Backpackers have travelled from being a little understood and fairly marginal segment of international tourism, to there being specific journal issues, new books and entire international conferences dedicated to the subject, with a new generation of PhD scholars in several countries working on many aspects (and this book contains some exemplary work from these new scholars). However, I would suggest that the literature, although now far wider (comprising both more case studies and newer aspects such as the anthropology and social construction of backpackers) needs a further deepening of analysis. To end this chapter, I would like to suggest that the journey of the emerging area of backpacker studies could be further enhanced by examining what could be called three 'spaces' on the journey within the growing study of backpackers across the world: the 'geographical space', the 'temporal space' and the 'political space'.

I would suggest that the first 'space' worth exploring is what could be called the 'geographical space'. In other words, I would argue that there is an overdue need for comparative studies of backpacker tourism across different regions of the world. Until now, backpacker research has tended to be clustered in certain areas, with a predominance of work based on South-East Asia and Australasia, and little published on Latin America, Europe and elsewhere. For instance, the useful work of Rogerson (2007) on South Africa's experience of backpacker tourism could be helpfully compared with the larger body of literature about South-East Asia, Australasia or Latin America. Specifically, this comparative process of working within a spatial aspect might raise questions, such as are there common themes that can be identified across regions, or are the regional differences too wide? Is historical specificity a key factor or not?

The second space is what could be called the 'temporal space'. Here I mean that there is also a need for longitudinal studies over time rather than the normal 'one-shot' type research that produces useful case studies, but tends to be a snapshot of a particular destination in time. By definition, such studies tend to lack the longer perspective of how destinations change and evolve over time. Also for the backpackers themselves, there is the important and not well-understood issue of their own travel cycles. Specifically, do they later return to the countries they visited as a backpacker, with their young families? If so, where do they now tend to stay, where do they now eat? Do they take their children to eat at the night market stalls or other small places or, do they tend to stay within the tourist bubble of the international hotels and resorts? Also, perhaps more seriously, it is not at all clear at present whether or not, overall, the backpacker experience has actually made any real difference to them or their values, lifestyle or other choices, or was it just a gap year before they rejoined the normal existence of work and career, starting a family, getting a mortgage etc.?

The third space, what I call here 'political space', concerns the thorny issues of power and power relations, that is, in essence the question of who wins and who loses from hosting backpacker tourism, particularly in LDCs. These are hard questions to ask, especially since the majority of tourism research tends to be case-study based. Asking such questions in many LDCs can be politically sensitive, and even asking such blunt questions may lead to problems of future access, visas being refused etc. for some outside researchers if they are seen to ask questions that are too pointed for local officials. Such a political economy-type approach would spotlight issues such as the local state level versus federal or national level planning tensions apparent in many LDCs concerning backpacker resorts. This could build on the work of Britton (1991) and Scheyvens (2002).

Finally, I would like to end by suggesting three specific research questions that could be usefully explored further, concerning enclaves, diffusion of taste and Asian backpackers. Although some excellent new work is appearing on backpacker enclaves (Lloyd, 2003; Howard, 2007; Brenner & Fricke, 2007), it is unclear what the exact role is of backpackers in the evolution of destinations, and specifically, are there differences between say present backpacker enclaves and other enclavic forms of tourism such as say, ecotourists?

In this chapter, I have attempted to self-consciously 'put myself back into the picture'. By doing this, I have found myself outside my comfort zone as an academic. I have experienced an intriguing tension created between the pull of trying to write this new story; and a pull the other way arising from my training and experience as a social scientist and of being immersed in the ways of being objective and in the need to remove myself from my writings and from the subject itself. In other words, unlike the more conventional papers that I normally write and publish, I have tried here to actively reflect on my own changing experience of the phenomenon of backpacker tourism, first as a backpacker myself in the days before the term really existed, through the 1990s boom in South-East Asia, to my return as an academic tourist studying backpackers. I have tried to interweave my own experiences with how both the backpacker phenomenon, and the academic study that it has generated, have evolved over time. As a former backpacker, and enthusiastic traveller (backpacker plus? post-backpacker? flashpacker even?) it is my hope that the broader narrative, the journey, continues. Intriguingly, there is some sense that the wheel has now come full circle, since my own story began with a journey to India in the 1980s. However, to end I would like to quote again from a recent post by Hutynk (2008) on his blog that 'Over twenty years it is common that youthful enthusiasms are tempered by the realisation that one ever knows less and less as knowledge grows'.

*Chapter 3*

# Flashpacking in Fiji: Reframing the 'Global Nomad' in a Developing Destination

JEFF JARVIS and VICTORIA PEEL

## Introduction

As a corollary of the rising mobility of youth, a growing multiplicity of tourist types choose travel experiences that reflect a desire to transform the space and time between key life events in diverse ways (Graburn, 1983; Sorenson, 2003). Sharing independent traveller routes with the stereotypical student backpacker in their early twenties are increasing numbers of both teenage 'gap year' travellers between school and university, and career break travellers aged from their mid twenties into middle life.

In 2006, during data collection for a larger study of the travel patterns of the backpacker/independent traveller in Fiji, the researchers encountered many of the latter who were stimulated to leave career jobs in professional fields to spend an extended period of time travelling or combining travel and work overseas. Such respondents often remarked that they had the financial capacity to stay at four or five star properties but instead chose to travel as a backpacker/independent traveller primarily due to the social nature of such travel. One single female traveller in her early thirties commented to the researchers that she checked out of a five star property in Fiji after one night as 'no one would talk to me there apart from the barman'. These typically older and more established travellers form a significant cohort within the larger backpacker taxonomy and are commonly referred to in the travel marketplace and media as 'flashpackers'.

While the term 'gappers' is understood primarily to mean younger, school or university break travellers, the term flashpackers frequently identifies travellers with higher disposable income and concomitant levels of expenditure. A consumer report conducted by The Future Laboratory in 2004 identified flashpackers as older travellers on career breaks who 'can afford to splash out on some of life's luxuries when the going on the road gets tough' (The Future Laboratory, 2004: 13). For others in the industry, the flashpacker is best identified by the expensive technology they carry with them while travelling: 'The flashpacker is a new breed of traveller, tech-savvy adventurers who have traded in their copy of "On the Road" for a cell phone, digital camera, iPod, wearable electronics clothes and a laptop, all snugly tucked away in their ergonomically correct, multi-function backpack' (*Breaking Travel News*, 2006). The media has also picked up on the term and uses it to describe 'upmarket backpacking'. For example, *The Guardian* newspaper in the UK ran a specific story on the flashpacking phenomenon in Fiji.

> If sleeping in a dorm with strangers snoring all night is no longer something you wish to endure, perhaps its time to admit that your backpacker days are numbered. You still want care-free days of river-rafting, visiting villages, buying local snacks, learning about the culture and meeting people, but you also need your own room with fluffy towels, a king size bed and designer toiletries at the end of each day? Then "flashpacking" is for you. (Miles, 2004)

A similar article in the *Sydney Morning Herald* quoted an industry operator attributing the rise of flashpacking to the increased number of older travellers on the road.

> We definitely see lots of older people, particularly 30 to 40, who are doing flashpacking around the world. One trend is that they have been backpackers before, but they are over the sharing of dorms, but they still go to backpackers. (Swart, 2006)

The development of an increasing number of well-appointed and often purpose-built accommodation, designed for this segment and marketed as flashpacker establishments is testament to their emergence as a sub-segment and their significance for the backpacker industry (Pursall, 2005). From the supply side, the media as well as many web sites discuss the rise of this flashpacker-style accommodation across the globe:

> Hostels have been popping up all over the world for the past 5 or 10 years with hotel-standard facilities at backpacker prices and these places seem to know what flashpackers want and are providing spotless hostels, comfy beds, and loads of facilities like bars, swimming pools, spas and all that good stuff. (Travoholic.com, 2006)

This chapter highlights the largely unexplored, emerging flashpacker sub-segment of backpacker tourism and examines the ways in which their travel behaviour and expenditure patterns presents a niche opportunity for sustainable tourism development in the Fiji Islands. Modelling the characteristics of the flashpacker identified by the popular media and tourism industry as discussed above, the study segments a survey of 696 independent travellers conducted in Fiji in 2006 in three ways: those who spent more during their stay in Fiji, those aged 28 years and over, and those who were travelling on a 'career break'.

In *The Global Nomad* (2004a), Cohen cautioned against research that homogenised the backpacker, calling instead for analysis into the diversity of the phenomenon specifically related to age, gender, sub-cultures and nationality. Likewise, Ateljevic and Doorne (2004: 76) noted that changes in the culture of backpacking in recent years confirms a 'continuing need for market research that reveals even more hetero-geneity and is context-specific'. Recent studies have moved to meet this challenge in identifying differences in backpacker motivation and experience specifically with regard to gender (Maoz, 2007, 2008; Muzaini, 2006) and nationality (Noy & Cohen, 2005), thereby contributing a more nuanced understanding of backpacker cause and effect at diverse destinations. Yet, while significant research has identified traveller age as a demographic factor in analysis (Jarvis, 2004; Scheyvens, 2006; Lee & Ghazali, 2008; Niggel & Benson, 2008), the behaviours and motivations of older backpackers in particular remain infrequently acknowledged or understood (Ryan & Mohsin, 2001; Speed, 2008). Myers and Hannam's (2008) study of women backpackers and Maoz's (2008) examination of middle-aged Israeli women backpackers are notable exceptions in their focus on travellers at diverse life stages.

References in the academic literature to the flashpacker phenomenon are concurrently rare, although interpretations of gappers, or gap-year travellers, are more established (O'Reilly, 2006). In an unpublished paper delivered to the Backpacker Research Group of the Association for Tourism and Leisure Education, Pursall (2005) commented on the evolution of a distinct flashpacker strand within the larger backpacker segment. His preliminary observations of traveller discussion boards noted how flashpackers differed from their backpacker cousins in their choice of more expensive accommodation and general disdain for the excessive revelry of frequently younger travellers (Pursall, 2005). In this context, flashpackers could therefore be considered higher yielding backpackers who can have a significant economic impact on the development patterns of host destinations.

## Backpackers, Flashpackers and Sustainable Tourism Development

From the mid-1990s onward, an increasing number of researchers have identified the benefits of backpacker tourism for the development of destinations despite limited support from host-destination governments. In his study on community tourism development in the Indonesian islands of Gilli Trawangan, Hampton (1998: 640) noted that backpackers are 'at best tacitly ignored, or at worst actively discouraged in official tourism planning'. This judgment has been largely supported in the literature on backpackers and destination development, which finds that the significance and benefits of the independent travel market have been greatly underestimated by countries seeking to develop in tourism (Hampton, 1998; Scheyvens, 2002, 2006; Cooper *et al.*, 2004; Visser, 2004; Lee & Ghazali, 2008).

In general, national tourism organisations have demonstrated negative or ambivalent attitudes toward youth travellers, who are often still perceived as impecunious drifters and socially undesirable 'hippies'. Yet, recent research has determined that independent travellers are of diverse ages, middle class, college educated and often motivated to travel because they are at a juncture in life (Moshin & Ryan, 2003; O'Reilly, 2006; Maoz, 2008). Almost all backpackers have a fixed return date and usually travel for between 2.5 and 18 months (Sorrenson, 2003). Very few travel for longer and even fewer drift aimlessly, in part because visas, passports and family ties are restricting. Indeed, the backpacker group is now recognised as increasingly diverse in motivation and demographics. What then of the missed opportunities presented to developing countries of the backpacker and the allegedly higher yielding flashpacker sub-segment?

Several studies have demonstrated the sustainable tourism policy implications provided by the 'form' of travel undertaken by the generality of youth travellers known as 'backpackers'. Here the term 'form' borrows from the analytical construction defined by Uriely *et al.* (2002) in their study of Israeli travellers to describe the institutional arrangements and practices by which tourists organise their journey. Both Hampton's (1998) study in Indonesia and Visser's (2004) investigation of backpacker tourism in South Africa discerned that the nature of small-scale, locally owned tourism businesses may be seen as a form of 'pro-poor' tourism and can provide a useful component of local economic development strategies for poor communities. Their research concludes that the capital requirements for starting small-scale businesses aimed at backpackers are modest and therefore more accessible for local providers. Due to the fundamental infrastructure of backpacker hostels, it was found, firstly, that locally manufactured goods were frequently employed, leading to lower foreign exchange requirements and minimal import leakages. Secondly, the establishment of hostels tended to have lower entry costs and minimal capital requirements, as they were generally renovated or converted properties rather than those that were purpose built. Thirdly, backpackers were found to contribute to the exchange of foreign earnings across larger areas than conventional mass tourists, as they tended to visit regions and backpacker hostels in more isolated areas.

In case studies of Asia, Africa and Samoa, Scheyvens (2002a, 2002b, 2005) also argues that the local servicing of the international tourism market challenges foreign domination of tourism enterprises, which usually occurs in the context of package tourism in the developing world. Local people and products can meet the needs of backpackers because they do not demand luxury and locals can work together to form organizations that promote local tourism, giving the community greater power over the industry (Scheyvens, 2005). As well as observing the immediate economic benefits for host communities, she notes that the backpacker's interest in local cultures revitalises cultural practices. Such findings have direct relevance to the future development of backpacker tourism in the developing economies of the islands of Fiji and the wider Pacific.

Backpackers are also regarded as 'hardy' travellers less dissuaded to travel by perceived risks in the developing world. Cook (1990: 14 as cited in Sonmez, 1998) defines instability as when a government 'has been toppled, or is controlled by factions following a coup or where basic functional pre-requisites for social-order control are unstable and periodically disrupted'. Political instability in Fiji following the May 2000 coup resulted in a sharp decline in tourist numbers, supporting research suggesting that tourists are concerned about safety in a destination (Hitchcock, 2001). As Hall and O'Sullivan (1996) reflect, 'perceptions of political stability and safety are a pre-requisite for tourist visitation. Violent protests, social unrest, civil war, terrorist actions, the perceived violation of human rights, or even the mere threat of these activities can all serve to cause tourists to alter their travel behaviour'.

However, backpackers have been found to be more risk averse than other tourists. Risk in a destination is perceived by backpackers as exciting and so risk is increasingly sought in travel. Lepp and Gibson's (2000) study of perceptions of risk associated with international tourism concluded that those who identified themselves as explorers or drifters were found to perceive less risk than other tourist roles. Elsrud (2001) also argues that risk taking and thrill seeking are an integral part of backpacking and that if a destination is considered risky, backpackers are less likely to be deterred. In addition, Riley (1988) identified the role played by independent travellers in defining a destination as safe for others. In this way, backpackers are pioneers of new destinations and remote areas in established destinations and are regarded as likely to be the first market to return post-crisis.

## Tourism in Fiji

Tourism in the Asia-Pacific region has grown significantly in recent decades with the number of arrivals doubling between 1990 and 2000. Economic growth and cheaper airfares have stimulated both intraregional and interregional travel, transforming several destinations into budget traveller enclaves. In Fiji, tourism has generated more foreign earnings than any other industry sector and is considered uniquely robust, having recovered quickly following the military coup of May 2000 (Ministry of Information, Communications and Media Relations, 2005). The tourism industry is mainly run by the private sector and contributes approximately 25% to the GDP in one of the world's least developed countries. The United Nations Development Programme (UNDP)/Government poverty study revealed that every fourth household in Fiji is struggling below the poverty line (Ministry of Information, Communications and Media Relations, 2005). This estimate is considered to be much higher than in previous decades and may be because of the political and economic instabilities following the most recent political coup. Inevitably, the Fiji government considers tourism to have the best prospects for building a more sustainable economy and future job creation.

Yet, despite promising growth within the tourism sector, the Fijian industry experiences several hurdles to sustainable tourism development endemic in developing countries. Two decades ago, Britton (1982) identified Fiji as facing high levels of foreign ownership, which restricts local participation in tourism to wage labour or small retail and artisan activities. Like many developing countries, Fiji's tourism industry has been held captive by a dependency on foreign capital reinforced by historical colonial ties that are difficult to disengage. Between 1988 and 2000, 94% of tourism projects implemented were foreign owned or joint ventures risking the loss of tourism earnings from the economy (Levitt & McNally, 2003).

A 'Strategic Environmental Assessment (SEA) of Fiji's Tourism Development Plan' conducted by The World Wide Fund for Nature – South Pacific Programme (WWF-SPP) and Asian Development Bank (ADB) found that although tourism is providing considerable economic benefits to Fiji, more than 60% of the money coming in, leaks back out of the country (Levitt & McNally, 2003). The continued development of five star hotels requiring substantial foreign investment is contributing to this problem, leading to a situation where money is transferred out of the country leaving little or no benefits to the local communities. In contrast to this pattern of five star resort development, Fiji has also developed a burgeoning backpacker travel industry, primarily in response to its accessibility as a 'round the world' airfare stopover and increased low-cost carrier access to New Zealand and Australia.

Fiji is one of the few countries in the world that has commenced collecting data on backpacker travellers via their International Visitor Survey (IVS) (Ministry of Tourism, 2006). In 2005, more than 3800 interviews were conducted with international visitors prior to their departure from the country. As part of that survey, interviewees were asked to describe their travel style. The 2005 IVS estimated that 'self-defined' backpackers represented approximately 12% of all visitor arrivals to Fiji, and over 14% of total international visitor nights. The IVS identified that the most popular regions for backpackers to visit were Nadi, the Yasawa and Mammanuca islands, and the Coral Coast.

## Methodology

This study was conducted in two phases and investigates international backpackers and independent travellers travelling in Fiji between 9 February and 2 March 2006 (phase 1: low season) and between 22 June and 9 July 2006 (phase 2: high season). At this time, budget carriers from both Australia (Pacific Blue) and New Zealand (Freedom Air) had recently entered the Fijian marketplace. This facilitated increased access to the destination from the densely travelled east coast of Australia and New Zealand. Fiji is also accessible as a popular 'round the world' airfare stopover for independent travellers from North America and Europe, as regular air services link Los Angeles and Auckland, Brisbane, Sydney and Melbourne.

Based on the 2005 Fiji International Visitor Survey data (Ministry of Tourism, 2006), key backpacker research destinations were identified within the areas of Nadi (the location of the airport and international gateway to Fiji), the Coral Coast and the Yasawa Islands. A cluster sample of popular backpacker/independent traveller accommodation facilities offering both dormitory rooms and private rooms or local Fijian *bures* were then identified through the *Lonely Planet Guide to Fiji* and informally via research with travellers during a pre-research visit conducted in 2005.

A number of popular accommodation locations were then selected as research hubs where research team members were based, these included the Nadi Bay Hotel (Nadi), The Beach House and Mango Bay (Coral Coast) and Oarsmans Bay and Mantaray Island (Yasawa Islands). The researchers also visited other backpacker accommodations within the vicinity of these research hubs, with those in the Yasawa islands accessible only by boat. In addition, research was also conducted on the only ferry connecting Nadi to the Yasawa islands (The Yasawa Flyer). At each site, random judgement sampling was used to select respondents based on the choice of the researcher. Blanket research coverage of all guests at the resort was carried out where possible and, overall, the researchers found very few refusals.

The method used for primary data collection was a face-to-face quantitative survey administered through personal interview and a non-probability sampling technique was applied. The data collection process relied on 'immersion methodology', where the researchers travelled as backpackers and stayed in travellers' accommodation. This methodology also permitted personal observation of the study group and informal conversations with the travellers typically during meal times. Suitable respondents comprised those who were staying in accommodation that offered bures/individual rooms as well as dormi-tory-style accommodation and who answered 'yes' to all screening questions. These addressed: having an independently organized travel schedule, time spent travelling away from home of at least a week, residency in a country other than Fiji and time spent in Fiji of at least one night (to exclude those just stopping over to change planes). The survey was conducted in two phases, one phase in high season and the other in low season due to time constraints associated with the availability of the research team. At the conclusion of the collection periods, data was collated and analysed using the SPSS statistical package.

In total, 696 usable responses were compiled. The sample was then segmented in three ways to identify potential flashpackers based on common media and industry interpretations of the travellers as high spending, older travellers and on career breaks.

Firstly, those who were self-budgeting to spend more than FJ$1499 during their stay in Fiji. In total 29.7% of the sample or 207 interviews were identified within the segment. For the purpose of this research, these high-yielding travellers were called flashpackers-HY.

The second segmentation of the sample focused on age. Those identified as 28 years or older were categorised as flashpackers 28+ ($n = 150$).

The third segmentation identified how travellers found the time to travel. Those who had previously left full-time employment and taken a break from a professional career were termed 'career breakers' or flashpackers-CB, in total 240 or 34.5% of the respondents fitted this description.

## Research Findings

### Age and nationality

As identified in Table 3.1, both flashpackers-HY and flashpackers-CB in this study were significantly older than the total results. Flashpackers-HY aged over 30 comprised 21.3% compared with 13.1% of the total sample, while flashpackers-HY aged under 20 comprised just 7.2% of the cohort compared with 14.9% of the full sample. However, it is also noteworthy that while 42.6% of high-yielding travellers were aged 26 and over, 57.4% of flashpackers-HY in this study were aged 25 years and under with 7.2% of these aged less than 20 indicating that younger travellers should not be stereotyped as low yield for destinations. With reference to flashpackers-CB, 54.8% were aged 26 or over in comparison to 30.6% of the total sample with an average age of 28. These findings confirm that flashpackers can be described as typically older travellers.

With regard to nationality, a total of 22 different countries were represented in the sample; however, Table 3.2 indicates that travellers

**Table 3.1** Age range

| Age (years) | Flashpackers-HY (%) (n =207) | Flashpackers-CB (n =240) | All travellers (%) (n =696) |
|---|---|---|---|
| Under 20 | 7.2 | 0 | 14.9 |
| 20–25 | 50.2 | 45.2 | 54.5 |
| 26–30 | 21.2 | 32.6 | 17.5 |
| Over 30 | 21.4 | 22.2 | 13.1 |
| Average age | 27.03 | 28.10 | 25.04 |

**Table 3.2** Dominant nationality groups

| Nationality | Flashpackers-HY (%) (n=207) | Flashpackers 28+ (%) (n=150) | Flashpackers-CB (n=240) | All travellers (n=696) |
|---|---|---|---|---|
| UK | 43.0 | 35.3 | 57.5 | 44.4 |
| Scandinavia | 15.5 | 10.0 | 5.8 | 11.3 |
| Australia | 11.6 | 11.3 | 1.3 | 7.6 |
| USA | 8.7 | 14.0 | 5.0 | 13.4 |
| Canada | 5.3 | 2.7 | 5.0 | 3.4 |
| Germany | 4.8 | 8.0 | 4.2 | 5.7 |
| Ireland | 1.4 | 4.0 | 11.3 | 4.7 |
| Other | 9.7 | 14.7 | 9.9 | 9.5 |

from the UK were dominant in the total sample (44.4%). Other leading national groups comprised the USA (13.4%), Scandinavia (11.3%), Australia (7.6%) and Germany (5.7%). The flashpackers-HY segment was dominated by British travellers; however Australians and Scandinavians had a higher representation in this segment than in the overall sample. The older flashpackers 28+ segment had the highest proportion of Americans at 14%, with Germans, Scandinavians and Australians also well represented. Significantly, flashpackers-CB travellers were highly concentrated from the UK (57.5%) and Ireland (11.3%) with very few Australian career breakers travelling in Fiji. The availability and popularity of the working holiday-maker visa in both Australia and New Zealand with young British and Irish travellers aged between 18 and 30 could be a reason for such a concentration, as Fiji is a popular stop over to and from both countries (Clarke, 2004).

**Travel style**

Respondents were asked to define their travel style during their trip to Fiji. As indicated in Table 3.3, just under half of the total sample perceived themselves as 'backpackers' (46.5%), just under one quarter identified with the term 'independent traveller' (24.0%) and over one tenth self defined as a 'tourist' (12.1%). The overall pattern remains the same for the flashpacker-HY sub-group, however, as they are more likely to identify as an independent traveller (26.6%), rather than a

**Table 3.3** Self-definition of travel style

| Travel style | Flashpackers-HY (%) (n =207) | Flashpacker 28+ (%) (n =150) | Flashpacker-CB (%) (n =240) | All travellers (n =696) |
|---|---|---|---|---|
| Backpacker | 41.5 | 22.0 | 48.3 | 46.5 |
| Independent traveller | 26.6 | 50.0 | 22.1 | 24.0 |
| Tourist | 14.5 | 10.0 | 10.0 | 12.1 |
| Backpacker/ independent traveller | 4.8 | 9.3 | 6.7 | 5.5 |
| Backpacker/ tourist | 3.9 | 1.3 | 5.0 | 5.2 |
| Independent traveller/ tourist | 2.4 | 4.0 | 4.2 | 2.7 |
| All three | 6.3 | 3.4 | 3.7 | 4.0 |

backpacker. It is noteworthy that travellers are inclined to eschew rigid definitions of their travel style, as a significant number of respondents identified variously as tourist, traveller, backpacker or any combination of the three.

The flashpackers 28+ group demonstrates significant differences from the other two sub-groups in that half the cohort identify as an independent traveller (50%) and only one fifth as a backpacker (22.0%). This finding duplicates that of Richards and Wilson (2004b) and suggests that the term 'backpacker' may be obsolete for this segment of older travellers and that those who are travelling in the style of a backpacker are increasingly dismissing the phrase.

Rejection of the backpacker term has significant implications for marketing Fiji as a backpacker/independent traveller destination. Personal comments from travellers indicated that the term 'backpacker' was increasingly perceived as associated with young travellers in their early twenties or late teens, who tended to travel in large groups, spend too much time carousing and too little time interacting with the local culture. The finding concurred with that of Doorne and Ateljevic (2005: 191), who observed that younger travellers are viewed by the veterans as 'inexperienced, commonly favouring "packaged" backpacker products which feature party culture and group activities' and Welk's (2005) description of the 'anti-backpacker backpacker'. Avoidance of the term by older travellers could also be in response to the recent growth in popularity of 18 to 19-year-old 'gap' backpackers travelling between school and university.

## Expenditure patterns

Travellers were asked to identify how much money they spent 'yesterday' on a variety of items. As the majority of surveys were conducted in the middle or later stages of their trip, it can be expected that transport costs were underestimated as travellers would often purchase bus or ferry passes valid for their stay at the start of their journey. For example, at the time of writing, a seven-day Yasawa Islands 'Bula' ferry pass cost FJ$279, while a two-week pass retailed at FJ$389 (www.awesomefiji.com, 2008).

In Table 3.4, travellers' daily expenditure demonstrates the particularly high-spending characteristic of flashpackers 28+ as a sub-segment. These older visitors spend an average of FJ$130.57 per day, which is FJ$24.88 or 23.5% per day above the average spend of FJ$105.69 for 'all travellers'. They tend to spend significantly more on accommodation (FJ$63.97), food and drink (FJ$33.83) and tours (FJ$10.84). Flashpackers-CB, who are absent from home for longer, have an expenditure pattern just FJ$9.05 more per day than all travellers.

**Table 3.4** Expenditure patterns

| Item of expenditure | Flashpackers-HY (FJ$) | Flashpackers 28+ (FJ$) | Flashpackers-CB (FJ$) (n=240) | All travellers (FJ$) (n=696) |
|---|---|---|---|---|
| Accommodation | 53.23 | 63.97 | 49.80 | 47.14 |
| Food and drink | 34.92 | 33.83 | 28.07 | 28.62 |
| Transport | 15.84 | 13.95 | 23.19 | 15.89 |
| Tours | 7.53 | 10.84 | 4.78 | 6.00 |
| Shopping | 4.63 | 5.31 | 3.79 | 4.47 |
| Entertainment | 2.58 | 2.67 | 5.11 | 3.57 |
| Total | 118.73 | 130.57 | 114.74 | 105.69 |

Flashpackers-HY are on the road for an average of 160 days and spend 17.1 of those days in Fiji, which accounts for 10.7% of their total time away from home. With an average expenditure of FJ$118.73 per day, each flashpacker-HY is therefore worth FJ$2030 to the destination. They also spend FJ$13.04 more per day than the average for all travellers in the sample. Flashpackers 28+ are on the road for an average of 152 days and spend 13.7 of those days in Fiji, which accounts for 9% of their total time away from home. With their average expenditure of FJ$130, each flashpacker 28+ is estimated to be worth FJ$1788 to Fiji. Flashpackers-CB are on the road for significantly longer, an average of 224 days, spending 14 days in Fiji and are estimated to be worth FJ$1606 per visitor.

## Motivation to Travel as a Backpacker or Independent Traveller

Travellers were asked to rate various travel motivations from 1 to 5. Table 3.5 demonstrates that all three groups of flashpackers exhibit similar patterns in motivation to all backpackers/independent travellers. Each identified taking 'an extended break from life at home' as the most important incentive, followed by 'a good way to meet other travellers', the low cost of or value for money nature of backpacking and the ability via backpacking to experience more of the country, culture and meet the locals.

## How Travellers Found the Time to Travel

Travellers were asked 'how they found the time to travel' on this trip to Fiji. Table 3.6 identifies the particular stage of life of the cohort and identifies those who are 'career gappers', taking an extended break between jobs, 'student gappers', taking an extended break during either their university degree or immediately after completing it (university gappers) or travelling between completing school and starting university (school gappers), on 'holidays' while in current employment, and retired. Significantly, 34.5% of the total sample were on career breaks.

Flashpackers-HY are more likely to be career gappers taking an extended break between jobs (42.0%) or on holidays from their current employment (18.4%) than the total sample. Flashpackers 28+ group are also more likely to be on a career break or paid holidays (61.3 and 27.3%, respectively). The propensity of flashpackers-HY to be university and high school student gappers (19 and 10.6%, respectively), and to be travelling while undertaking a university course in Australia or New Zealand (9.2%), is also notable.

**Table 3.5** Motivation to travel as a backpacker/independent traveller (score out of 1–5)

| Motivation | Flashpackers-HY | Flashpackers 28+ | Flashpackers-CB (n =240) | All travellers (n =696) |
|---|---|---|---|---|
| To take an extended break from life at home | 4.1 | 4.2 | 4.3 | 4.1 |
| It is a good way to meet other travellers | 4.0 | 4.1 | 4.1 | 4.0 |
| It is a cheaper way to travel | 4.0 | 4.0 | 4.1 | 4.1 |
| To experience more of the country | 3.9 | 4.0 | 3.9 | 3.9 |
| It is a good way to experience Fijian culture | 3.8 | 4.0 | 3.8 | 3.8 |
| It is a good way to meet local people | 3.8 | 3.8 | 3.7 | 3.8 |
| Travelling is a good party atmosphere | 3.2 | 3.5 | 3.2 | 3.2 |
| It is economically beneficial for the local community | 3.1 | 3.2 | 3.3 | 3.2 |
| It is an environmentally sustainable way of travelling | 2.9 | 2.8 | 3.0 | 3.0 |

## Discussion

### Who is the flashpacker?

In seeking to identify the flashpacker as represented in the island resorts of Fiji, a number of observations can be made from this study by segmenting the total sample in different ways. The high yield sub-segment flashpackers-HY who budgeted to spend more than FJ$1499 in Fiji are more likely to be older travellers in their mid to late 20s and beyond, many with professional careers who are travelling on a career break or extended holiday between employment, although significantly,

**Table 3.6** How travellers found the time to travel

| Life stage | Flashpackers-HY | Flashpackers 28+ | All travellers (n=696) |
|---|---|---|---|
| Career gapper | 42.0 | 61.3 | 34.5 |
| Student gapper (university) | 19.3 | 4.0 | 22.4 |
| Paid holidays | 18.4 | 27.3 | 13.8 |
| Student gapper (school) | 10.6 | 0.0 | 18.5 |
| Study tourist (at university in Australia – New Zealand) | 9.2 | 5.4 | 10.4 |
| Retired | 0.5 | 2.0 | 0.4 |

younger travellers are also represented. They tend to associate more with the term 'independent traveller' rather than 'backpacker', but are cognizant of the ways in which their travel at different places and at diverse times is akin to what they understand to be that of a backpacker, independent traveller or tourist.

The older sub-segment, flashpackers 28+, spent an average FJ$130.57 per day in Fiji, which equates to 23.5% more per day than the average of all travellers. They are mainly professionals on a career break or extended holiday, who chiefly describe themselves as independent travellers with under one quarter (22%) self-defining as a backpacker. As expected, flashpackers-CB are also older travellers with 54.8% represented as aged 26 and over who travel on the road for longer than the other two sub-segments. They are more likely to be from the UK and Ireland with only 48.3% self-defining as a 'backpacker'.

The analysis of the three different traveller segmentations conducted in Fiji suggests that, overall, flashpackers travel for similar motivations to other travellers and use the same travel routes and infrastructure existing for all independent travellers. However, as they can be regarded as older travellers, their access to higher disposable incomes means they can demand elevated standards of accommodation and tend to spend more per day on meals and tours.

While flashpackers and the motivations and behaviours of older, career-break travellers demands further analysis, this study posits a tentative definition of the group, adding the significant dimensions of life stage and 'how they found the time to travel' to the profile. Building on Pearce's (1990) seminal classification of the late 20th-century backpacker, a flashpacker exhibits all of the following characteristics:

- a traveller who found the time to travel by either being on a career break or an extended holiday from paid employment;
- typically aged in their mid twenties and upward;
- a preference for small scale, value for money (not necessarily budget) accommodations;
- an emphasis on meeting other travellers and locals (where possible);
- an independently organized and flexible travel schedule;
- a preference for longer rather than brief holidays (where possible);
- an emphasis on informal and participatory holiday activities.

The emergence of flashpacking is due to a number of significant contemporary drivers. In discussing issues in modern work-life balance, *The Economist* magazine observes that workers today are happy to 'binge-work for a while, but in return want extended sabbaticals in which to chill out' (*The Economist*, 2006: 78). In a new age of hyper-mobility (Richards & Wilson, 2004a: 3; Molz, 2005), the career break phenomenon has developed to address diverse social and economic challenges. Together with the contemporary conundrum of work-life balance, the postponement of family responsibilities and the changing realities of the workplace where limited contract positions and career/job switching is common, are both contributing factors in the evolution of the contemporary flashpacker. Inevitably, the tourism industry is responding with the development of specific internet sites promoting career breaks and discussing travel options, such as www.thecareerbreaksite.com. Here, potential career breakers are described in the following manner in response to the question, 'who takes career breaks'?

> they tend to be aged 25–34 and from a professional background. Many career breakers are at the point where they're going to change jobs, and want to take some time out before they start working again. A lot of people go travelling after a relationship has broken up. (The Career Break Site, 2008)

In assessing the human consequences of globalisation, Bauman (1998) identified the emergence of a professional and managerial elite who service and maintain the new global economic system. He asserts that hyper-mobility is now automatic in both the working and personal lives of this elite, where travel enables freedom on an unprecedented scale. Taking a career break and specifically working for a period professionally in another country can also be seen in the context of globalisation as an investment in the future career of the traveller. O'Reilly identified that the argument that 'the trip will be beneficial for a future career' is a common 'excuse' that is often sold to 'significant others' (O'Reilly, 2005). Pearce and Foster (2007) specifically identified that social interaction skills, self-management skills, social and cultural awareness, independence, dealing with difficulties, self-confidence and problem solving, were all skills that the travellers had developed while travelling and believed were important for future employment opportunities.

Flashpacking has evolved as a result of diverse societal factors, including the trend to cosmopolitanism, which 'involves the search for, and the delight in, the contrasts between societies' (Szerszynski & Urry: 467). Urry (1995) identified 'the cosmopolitan person' as a highly mobile, curious, open and reflexive subject who delights in and desires to consume difference and, in doing so, acquires the competence to navigate in an increasingly diverse global environment. Such a description parallels the motivation to seek cultural immersion at the host destination identified as central to backpacking ideology (Riley, 1988).

## Flashpackers and Sustainable Tourism in Fiji

For destinations along independent traveller routes around the world, such diverse patterns of mobility are highly significant (Urry, 2000). The emergence of the flashpacker segment of the broader backpacker/independent traveller market can address specific problems related to the sustainable development of Fiji's tourism economy such as; low yield per visitor, economic leakage, indigenous participation in the industry and perceptions of Fiji as a risky destination. Leakage occurs when imports are used to satisfy demand and when developments are foreign owned. By contrast, small-scale tourism developments, such as those serving the independent traveller market, have far lower rates of economic leakage.

The Fijian independent traveller/backpacker industry, which is becoming increasingly centred on small-scale resort development in the Yasawa Islands, bears an inherent advantage over other developing destinations in that the majority of businesses are indigenously owned and operated by employees from the local villages. This is partially due to the administrative difficulty that foreigners have in leasing land in the island chain, as well as a strong interest in the development of small-scale tourism from local villagers. At the time the research was conducted in 2006, only one backpacker resort was owned by foreigners (a Swedish–Australian venture), who were required to employ the local villagers as a consequence of their lease. Recent growth in the number of 'integrated properties' providing both dormitory and more expensive private beachfront bures suggests nascent recognition of the positive attributes of a diversified backpacker market. While flashpackers are more demanding of comfort and privacy than backpackers, the cost of developing accommodation suitable for the market is likely to remain financially viable for local communities.

An interview conducted by the researchers with an indigenous backpacker resort owner in the Yasawa islands estimated that the construction costs, using local village labour, of a Fijian-style bure were estimated at FJ$500–800. After completion, the same bure was rented out to independent travellers at FJ$110 per night. Thus, within a week or so of occupancy, the net construction costs of the bure are recovered. Other resorts in the Yasawas provided more basic 'western-style' weatherboard and wooden huts with tin roofing, balconies facing the beach with fans and some with en suites; these were more expensive to construct, with figures quoted to the researchers varying from the FJ$8000 to FJ$12,000 mark, depending on the level of luxury. These were then rented at a charge of approximately FJ$200–250 per night; a figure which would still see construction costs recouped in under two months of occupancy. In addition to recovery of costs on accommodation, the survey highlighted that flashpackers-HY spent significantly on food and drink, and this expenditure increased for the flashpackers 28+ group. Considering that there are limited dining options in many parts of Fiji, especially in the Yasawas, the accommodation sector that supplies meals for backpackers and flashpackers may be gaining as much 73% of the revenue.

## Conclusion

In this chapter, life-cycle diversity and the emergence of the flashpacker sub-segment within the current independent traveller population is understood as both an inevitable result of globalisation and the increased mobility of youth, and as a source of possible advantage for community tourism engagement in a developing destination such as Fiji. Flashpackers have been defined here as independent travellers aged from their mid twenties upwards who are travelling on career breaks or on extended paid holidays. Significantly, as older travellers, they are less likely to self-define as backpackers, although they are motivated to travel in similar ways to the backpacker and use the same traveller infrastructure. By contrast, however, they have the ability to spend more per night than other travellers and therefore present a significant earning opportunity for developing destinations. It is incumbent upon policy makers within host destinations such as Fiji to recognise the diversity within the previously homogenised market segment of backpacking and to find ways of supporting the local industry in addressing the new demands associated with the emergence of flashpacking.

Chapter 4

# The Virtualization of Backpacker Culture: Virtual Mooring, Sustained Interactions and Enhanced Mobilities

CODY PARIS

## Introduction

Backpacking is a culture symbolic of the increasingly mobile world. With cultural roots growing from the beatnik and hippie countercultures of the 1950s to the 1970s, backpacking has been a mainstreaming phenomenon in tourism that has evolved and adapted to technological, social, political and economic trends in both the home and destination societies of backpackers. These global trends over the last 50 years have led to a democratization of backpacking to a large, heterogeneous and globally diverse group of people (Paris, 2008). The pillars of ideology of the backpacking subculture (Welk, 2004) have persevered over the last few decades, but the social cohesiveness, imparted early on by the close connection with the social countercultures of the time, arguably, has not. One of the largest constraints to depicting backpacking as a subculture is that it has become so mainstreamed (Scheyvens, 2002; Welk, 2004; O'Reilly, 2006); it is difficult to illustrate clearly the boundaries between the backpacker community and other mainstreamed tourists. Welk (2004) reasoned that since backpacking has lost the countercultural connection of its roots, for today's backpackers, backpacking is just a short-term countercultural experience along a set 'backpacker trail', and that they are re-assimilated into their home society upon returning.

Recent developments in information and communications technology have provided the basis for the backpacker culture to, once more, gain the cohesiveness without the temporal or spatial constraints of the 'backpacker trail'. While the physical mobilities of backpackers are still just as important to the backpacking experience, new virtual moorings

(Hannam *et al.*, 2006; Ateljevic & Hannam, 2008) have developed that allow backpackers to be fully integrated in their multiple networks and maintain a sustained state of co-presence between the backpacker culture and their home culture (Mascheroni, 2007). Backpackers manage their multiple networks both while travelling and at home, using social networking sites (SNSs), email and other technologies, which have simultaneously blurred the boundaries between home and away. The close virtual proximity that backpackers maintain allows them to be instantly in contact with friends, family, work, school and fellow travellers. Further, the backpacker ideals of independence, freedom and physical travel are all enhanced by the virtual mobility of backpacker information, identities and culture. The purpose of this study is to firstly explore the complex sociality backpackers now maintain by examining the convergence of the internet and communication technologies and physical travel before, during and after backpackers' trips, and secondly to develop an understanding of the stabilization and creation of the social structure of the backpacking culture that has resulted from the convergence of technology and ideology. Through this examination of the virtualization of the backpacker culture, this study makes the argument that the recent innovations of the internet and communications technologies have provided the social structure that allows the ideological system of the backpacker culture to be maintained on the road as well as at home.

## Theoretical Background

### Virtual mobility and backpacking

Recent innovations in the internet and communications technologies have created a more networked patterning of social life, home life and work life (Hannam *et al.*, 2006). These technologies have allowed many people to maintain intermittent co-presence within these networks. Co-presence is further enhanced by 'virtual travel', as many social interactions need to take place over long distances, where corporeal travel is not as easy. This virtual proximity is proliferated by advances in cyberspace, including email, SNSs, blogs and other virtual extensions of personal identity. The virtual proximity of an individual's multiple networks allows them to easily shift between or simultaneously interact with more than one network. In the increasingly complex world, where people need to maintain close networks over large geographical distances, virtual mobility allows for the strengthening of interactions (Urry, 2002a). The virtual mobility of personal networks allows people to connect to their networks anywhere and at anytime, especially with advances in personal wireless technologies (Hannam *et al.*, 2006). The spatial division between 'home and away' is now less important, allowing people greater flexibility with concern to their movements through time and space. Many new jobs allow people to work anywhere they have a connection and extended education programs allow people to receive knowledge over long distances. Moreover, the profusion of information available and adoption of e-commerce by travel service providers have made independent corporeal travel much easier. The understanding of the convergence of travel and communications technology can be enhanced through the examination of the current state of backpacking.

Mobility is inherently part of the backpacker phenomenon (Ateljevic & Hannam, 2008). Figure 4.1 presents the theoretical framework of backpackers' mobilities. The framework illustrates three 'spaces' of backpacking, and the intersections or mobilities between the spaces. The physical spaces include backpacker destinations, enclaves, hostels, specialized travel agencies, internet cafes, transportation, home locations

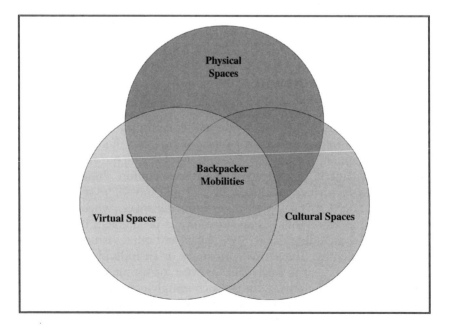

**Figure 4.1** Theoretical framework of backpacker mobilities

and other backpacker travelscapes (O'Regan, 2008). The cultural spaces represent the shared ideology, identity, social status, motivations and attitudes, 'outside' representations and perceptions of backpacking (Bennett, 2008) and transformative affects of backpacking. The virtual spaces of backpacking include email, online communities, blogs and personal websites, as well as mobile devices (laptops, mobile devices, cell phones) and connections (internet, wi-fi, broadband). The development of these technologies has provided the communication tools that allow travellers to stay in contact with friends, family and other travellers while away. While at home, the virtual mobility technologies have enabled travellers to share their experiences as well as virtually take part in other travellers' experiences. Backpackers have fully adapted the new communication tools into their travel activities, thus giving rise to a new virtual mobility. Travellers can now be in two places at once, their physical location either at home or on the road, and the virtual location. Mascheroni (2007) explored the convergence of new communication media, the internet and mobile phones, and travel by backpackers. Mascheroni (2007: 527) stated that 'global nomads produce and maintain mobile spaces of sociality, founded on a complex interaction of face-to-face interaction and mediated communication, co-presence and virtual proximity, corporeal travel and virtual mobilities'. Backpackers have increasingly used email and SNSs to stay in contact with fellow backpackers met during their trip, and their social network thus becomes accessible anywhere at anytime (Mascheroni, 2007). The increased number of backpackers that bring laptops or wi-fi-enabled cell phones and PDAs combined with the proliferation of the availability of wireless and land-line internet access in hostels and guesthouses, cafés, restaurants and bars in backpacker enclaves provide an almost constant connectivity to the internet. The developments in technology have been combined with other global developments including increasing global, middle class, socially acceptable travel (Gap Years, Overseas Experience, study abroad), career breaks, changing family and work obligations, advances in credit card technologies, etc. The list can continue, but all of these advancements mean that a new type of mobile traveller can exist now that could not 20 years ago. One that can be anywhere in the world, but still manage their multiple networks, simultaneously. O'Regan (2008) characterized backpackers as part of the new mobile elite, who are globally mobile, and are defined by their relationship to technology, financial capital and information.

The internet has been identified as a large source of just-in-time (Adkins & Grant, 2007) information for backpackers when planning, as well as a tool for booking travel, which allows backpackers to maintain the independent nature of their travel experience. The interaction with fellow travellers in online web communities has provided backpackers with a useful source of travel information, and possibly more important, it has provided a mobile structure in which the backpacker subculture can exist. Many backpackers use narrative emails and/or travelogues via personal websites, blogs or SNSs to share their experiences with friends and family back home, as well as fellow travellers. Mascharoni (2007) points out that travelogues allow backpackers to maintain mobile spaces of sociality that follow the individual during their trip, are constantly updated and with an email address that represents the only permanent address of the traveller. Mobile sociality allows backpackers to maintain connections with fluid social networks made up of friends, family, travellers met while on the road, local people and unknown travellers (Mascharoni, 2007). Adkins and Grant (2007) note that information and communication technologies develop solidarity and a culturally shared understanding of what a backpacker is, which could conflict with the commercial backpacker image. Constant contact with the subculture through the internet can help to cultivate an individual's backpacker identity, while at home or on the road (Paris, 2008; Mascheroni, 2007; Sorensen, 2003). The following section delves deeper into the understanding of backpacking as a cultural phenomenon and the implications of the convergence of travel and communications technology for backpackers.

## Changing Backpacker Culture

Many researchers have been hesitant to examine backpacking in its entirety as a culture (Anderskov, 2002), instead examining a part of backpacking, or categorizing backpacking as a scene, community, market, type, form or a tribe. Welk (2004) understands the backpacker community as a scene, or an informal social group with undefined boundaries. Members of a scene often share common leisure interests and frequent particular 'hangouts'. Welk also argues that the backpacker community is not a real subculture, but it does take on some of the characteristics of one constrained by a constant assimilative pressure by mainstream tourism. Backpackers have also been characterized as neo-tribes (Mafessoli, 1995; Wilson & Richards, 2004d), social groups that are temporarily bonded together. Backpackers on the road are temporary members of the 'road culture', participating in short-term fleeting social interactions, often travelling together, eating together and sharing common experiences together. The social interactions (Murphy, 2001) between them revolve around the shared ideology and a sense of companionship in the midst of the social insecurity of travelling in a distant unknown place, but are constrained to just the time spent while travelling. This sociality is maintained through a shared ideology.

Welk (2004) proposed that the backpacker community has evolved around a set of stable common symbolic lines of ideology. Welk suggests five pillars of *backpacker ideology* including: to travel on a low budget, to meet different people, to be free, independent and open-minded, to organize one's journey individually and independently, and to travel as long as possible. Yet, as a result of the global trends during the last few decades, many of the current characteristics of backpackers differ from those of the 'drifters' and 'nomads from affluence' (Cohen, 1972, 1973) of the 1960s and 1970s, but it is the ideology, motives and intentions that have persevered (Sorensen, 1992, 1999, 2003). The stability of the ideology continued through the 1990s, exemplified by the ethnographic study of the backpacker culture conducted by Anderskov (2002) in Central America, which found that the key elements of the backpacker culture, succinctly outlined by Welk's five pillars, were essentially the same as they were in Sorensen's (1992) study, even though the studies were conducted on two different continents a decade apart. Many backpackers' experience with the backpacker culture was limited to the 'road culture' (Sorensen, 2003) due to corporeal mobility, temporary social interactions, and a limited amount of time participating in the culture; consequently, few new practices were created that led to changes in the structure and values of backpacker culture.

Anderskov (2002) concluded that the backpacker culture is hierarchically structured, with individual status connected to the values of freedom, independence, tolerance, low budget and interaction with locals. Further, the author argued that social hierarchy of the backpackers was dependent upon the exchange of the most valuable object in the culture, information. While the ideology of the backpacker market is represented in the value system, there is a noticeable gap between the structure and the actual practice by backpackers (Anderskov, 2002; Richards & Wilson, 2004d). Both Sorensen (1992) and Anderskov (2002) note that backpacking culture is unique in comparison to other travel cultures in that most respondents ideally wanted to project the values gained as participants in the backpacker culture over to their 'normal' lives.

Technological innovations have contributed to evolutions and adaptations in the social systems of the backpacker culture. In the 1970s, the increase in the number of young independent travellers with a strong association to the counterculture, commonly referred to as drifters (Cohen, 1973), led to the development of the early backpacker trails and enclaves, including the 'hippie trail' (Cohen, 1973). It was after taking a long-term overland trip from Europe to Australia via Asia on the hippie trail that Tony Wheeler and his wife wrote the first *Lonely Planet* guidebook (Welk, 2008). Interdependently, the development of alternative guidebooks, like *Lonely Planet*, and the evolution of the social systems of backpackers occurred (Welk, 2008). The guidebook represented a fixed structure of the backpacker culture that facilitated information and culture transfer between current backpackers, new backpackers, and from one generation to another (Sorensen, 2003). Guidebooks provided a common reference source for backpackers, while solidifying the backpacker ideology of independent, free and long-term travel. Guidebooks also reinforced the development of backpacker trails and enclaves by providing information to cultural insiders. The expansion of the guidebooks contributed greatly to the mainstreaming of the backpacker culture, as they made the actual act of backpacking easier. Guidebooks are not the only reason for the expansion of backpacking. The development of an advanced infrastructure of backpacker accommodations, travel agencies and transportation all contributed greatly, as did other global trends, such as the increasing global middle class, increasing amounts of disposable income of younger people, increased social support for backpacking in the form of the Overland Experience (Bell, 2002) or the Gap Year (O'Reilly, 2006) and the increased representation of backpacking in pop culture (Richards & Wilson, 2004d). A complete examination of the backpacker culture subsystems would include evolutionary and adaptation analysis of all these trends. As noted in the introduction, the mainstreaming of backpacking is considered the main barrier to examining backpacking as a culture, or in other words, backpacking culture has outgrown the social structure needed to support it.

## Methodology

A self-administered online survey was used to collect the data for the study. Surveys were administered through specific online backpacker groups. A link to the survey and a short message explaining its purpose was posted on the discussion boards of 15 backpacker-specific groups on Facebook.com and on several threads of Lonely Planet's Thorn Tree Forum of which members had to choose to join. This methodology has proved to be effective in overcoming some of the problems associated with targeting an inherently mobile group, as is the case with backpackers (Paris, 2008).

The questionnaire was designed to measure the backpackers' demographic information, including age, gender, education, employment status, nationality and previous travel experience. The questionnaire also employed a set of questions developed by the author from an analysis of the literature, previous research experience and several informal interviews with backpackers about their online habits. Using a Likert scale, the questionnaire measured how the internet was used, perceived and represented by backpackers before, during and after their most recent trips. Data was collected during the one-month period between 15 August and 15 September 2008. The data analysis procedure included descriptive statistics, and analysis of variance (ANOVA) to examine the differences between the age cohorts.

## Results

### Sample profile

The sample profile is presented in Table 4.1. The gender breakdown is similar to that of the most recent quantitative studies on backpackers (Paris, 2008). While the majority of the sample is under 30 years old (65.5%), about one third of the sample is older than 30 years. Education level is considerably high, and about 20% of the respondents are currently students. Most of the sample is employed either full or part time, suggesting that the respondents would be temporally constrained if and when they take a trip. The nationalities of the sample are presented in Table 4.2. As expected, most of the respondents are from North America, Western Europe or Australia. However, there is an increasing amount of backpackers from 'non-traditional' source countries with about 8% of the respondents from Latin America, Malaysia, South Africa or another Asian country.

Respondents' previous travel experience is presented in Table 4.3, and is similar to the travel research presented in recent studies (Paris & Teye, 2009; Paris, 2008). Respondents had varying levels of travel experience. In terms of the number of countries visited by respondents, most had either visited between nine and 20 countries or more than 25. There is a similar pattern in the number of international trips taken by respondents, as most had taken less than 10 or more than 16. The largest percentage of respondents' previous trip lengths was between two and six weeks, suggesting that because of external constraints, many backpackers' trips

**Table 4.1** Sample profile

| Attribute | No. | % |
|---|---|---|
| **Gender** | | |
| Male | 94 | 43.3 |
| Female | 123 | 56.7 |
| **Age (years)** | | |
| 18–24 | 67 | 30.9 |
| 25–30 | 75 | 34.6 |
| 31–40 | 33 | 15.2 |
| 40+ | 42 | 19.1 |
| **Education** | | |
| High School (up to year 12) | 25 | 11.5 |
| College (4 year) | 96 | 60.8 |
| Graduate School (advanced degree) | 119 | 27.6 |
| **Employment** | | |
| Student | 45 | 20.7 |
| Employed full-time | 120 | 55.3 |
| Employed part-time | 27 | 12.4 |
| Unemployed | 25 | 11.5 |

are relatively 'short term' (Sorensen, 2003; Paris, 2008). Still, one fifth of the sample's previous trips lasted between three months and one year.

## Pre-trip results

Sample mean scores were calculated for responses to the pre-trip questions and are presented in Table 4.4. Respondents' information sources included both the internet and the traditional sources of alternative guidebooks. Information was gained virtually by visiting general information sites, other backpacker's personal blogs and websites, virtual communities and through virtual interactions with personal networks composed of friends, family and other travellers. More interactive sources of information, such as online video sources on

**Table 4.2** Nationality

| Nationality | No. | % |
|---|---|---|
| USA | 50 | 23.0 |
| UK | 40 | 18.4 |
| Canadian | 27 | 12.4 |
| Australian | 25 | 11.5 |
| Scandinavian | 16 | 7.4 |
| German | 15 | 6.9 |
| Other European | 13 | 6.0 |
| Latin American | 6 | 2.8 |
| Israeli | 5 | 2.3 |
| Irish | 5 | 2.3 |
| New Zealander | 4 | 1.8 |
| Malaysian | 4 | 1.8 |
| Other Asian | 4 | 1.8 |
| South African | 3 | 1.4 |

Youtube.com or virtual tours of accommodations or attractions, were not utilized as much. Many respondents also booked travel online, specifically hostel accommodations prior to departure.

Table 4.5 presents composite scores and the results of the ANOVA for the pre-trip questions between the four age cohorts. The pre-trip questions were concerned with three topics, interaction with other backpackers, information search and travel booking. The results suggest that online forums provide a strong source of both information and interaction between backpackers of all ages. Interestingly, the participation in online forums increased as age increased. While the 41+ age cohort was the most likely to have visited online forums, they were by far the least likely to join a backpacker-specific group on a SNS. This could be attributed to the young average age of most users of SNSs, however, the two age cohorts representing people between 25 and 40 were the most likely to join backpacker groups on SNS before departing. This suggests that while many of the members of the youngest cohort have yet

**Table 4.3** Respondents travel experience

| Travel experience | No. | % |
|---|---|---|
| **No. of countries visited** | | |
| 1–4 | 8 | 4.8 |
| 5–8 | 16 | 9.5 |
| 9–12 | 35 | 20.8 |
| 13–16 | 21 | 12.5 |
| 17–20 | 25 | 11.5 |
| 21–24 | 10 | 6.0 |
| 25–30 | 19 | 11.3 |
| > 30 | 34 | 20.2 |
| **No. of international trips** | | |
| 1–3 | 22 | 13.1 |
| 4–6 | 32 | 19.0 |
| 7–10 | 25 | 14.9 |
| 11–13 | 12 | 7.1 |
| 14–16 | 12 | 7.1 |
| > 16 | 65 | 38.7 |
| **Average length of previous trips** | | |
| 0–2 | 20 | 11.9 |
| 2–6 | 83 | 49.4 |
| 6–11 | 28 | 16.7 |
| 12–24 | 23 | 13.7 |
| 25–52 | 14 | 6.5 |

to have the 'on the road' experience with the backpacking culture, and there is no connection yet to maintain a virtual presence through the SNS groups. The pre-trip information search questions indicate that guidebooks are still very important for backpackers of all ages. The use of the internet for pre-trip information searches, however, seems to be more

**Table 4.4** Pre-trip questions

| Item | Mean (n =217) | S.D |
|---|---|---|
| Use the internet to search for information about the destinations you are visiting | 4.61 | 0.693 |
| Bought a guidebook like *Lonely Planet, Let's Go* or *Rough Guides* to help plan your trip | 4.38 | 1.007 |
| Have visited backpacker forums such as Lonely Planet's Thorn Tree Forums, or other online backpacker travel communities | 4.19 | 1.084 |
| Have sought out tips from friends, family and other travellers about your trip online | 3.79 | 1.232 |
| Book travel online before leaving home | 3.53 | 1.198 |
| Visited other backpacker's blogs and websites | 3.45 | 1.220 |
| Have booked a hostel on a website like Hostelworld.com or hostelbookers.com prior to leaving home | 3.10 | 1.536 |
| Have joined a backpacker-specific group on a social networking site | 2.78 | 1.495 |
| Took virtual tour of accommodations, attractions etc. | 2.63 | 1.348 |
| Visited websites like youtube.com to view videos or pictures about the destination you are travelling to | 2.62 | 1.304 |
| Have arranged to meet up during your trip with someone that you met online | 2.38 | 1.307 |

important, as the composite mean scores for all age cohorts were greater for the use of the internet for obtaining information than for the use of guidebooks. Respondents, during their information search, also visit other backpacker's online blogs and websites. The 25–40 age cohorts were the most likely to visit other backpackers blogs. Blogs provide backpackers with information and a medium for individuals to assert their own 'traveller' status. All of the respondents sought out tips from friends, family and other travellers online, with the 18–30 age cohorts having the highest median scores. This suggests that younger respondents are regularly connected to their virtual networks, consisting of their friends and family, as well as their peer travellers. The results also

**Table 4.5** Comparison of pre-trip questions between age groups

| Travel behavior | 18–24 | 25–30 | 31–40 | 41+ | F value | Sig. |
|---|---|---|---|---|---|---|
| Have visited backpacker forums such as Lonely Planet's Thorn Tree Forums, or other online backpacker travel communities | 3.89[1] | 4.23[12] | 4.39[12] | 4.60[2] | 4.709 | 0.003 |
| Have joined a backpacker-specific group on a social networking site | 2.96[1] | 3.09[1] | 3.12[1] | 1.74[2] | 9.642 | 0.000 |
| Use the internet to search for information about the destinations you are visiting | 4.48[1] | 4.65[12] | 4.85[2] | 4.62[12] | 2.371 | 0.071 |
| Visited other backpacker's blogs and websites | 3.36 | 3.56 | 3.85 | 3.14 | 2.402 | 0.069 |
| Have sought out tips from friends, family and other travellers about your trip online | 3.89[12] | 4.00[1] | 3.79[12] | 3.31[12] | 2.964 | 0.033 |
| Bought a guidebook like *Lonely Planet, Let's Go,* or *Rough Guides* to help plan your trip | 4.41 | 4.28 | 4.39 | 4.43 | 0.245 | 0.865 |
| Book travel online before leaving home | 3.49 | 3.42 | 3.55 | 3.71 | 0.511 | 0.675 |
| Have booked a hostel on a website like Hostelworld.com or hostelbookers.com prior to leaving home | 3.18 | 3.21 | 3.06 | 2.83 | 0.594 | 0.619 |

*Note:* Superscript numbers denote significant difference between means at $p < 0.05$. Tukey's HSD was used. Likert scale used (1 = low, 5 = high).

suggest that respondents of all ages book travel prior to departure, and that younger backpackers book hostels online prior to travel. The older cohorts tend to book hostels less prior to their departure, but this is likely a result of the accommodation choice of many older backpacker, who are less likely to stay in hostel-style accommodations (Paris, 2008).

## During-trip results

Composite scores for responses to the during-trip questionnaire are presented in Table 4.6. The questions in this section were concerned with the virtual proximity of backpackers to their networks, the tools used to maintain that virtual proximity and the use of virtual tools to enhance their corporeal mobility while travelling. Respondents maintained direct contact with friends and relatives primarily through email, but they also used SNS and traditional communications methods, i.e. postcards, calling cards or call centers, and to a lesser extent cellular phones. Respondents were less likely to maintain contact with work using cell phones or email while travelling, suggesting that even though the means to facilitate such a connection were available, backpackers chose not to. The findings also suggest that backpackers indirectly maintain contact with friends and family back home by posting photos online, and to a much lesser extent updating a personal blog. Many respondents agreed that while travelling, an individual's email address or social networking profile represents their only real address, which suggests that even though they are corporeally mobile, backpackers maintain intermittent co-presence with their networks through an immobile and virtual identity. The actual independence and freedom of movement while on the trip was enhanced by the backpackers' ability and willingness to connect to their virtual networks. Many backpackers agreed that they used the internet and, specifically, online backpacker forums to gain new information, contribute new information and to secure bookings for upcoming travel services. Also, backpackers sought to maintain the social interactions experienced while on the road by adding friends made while travelling, to their social networks online. In order to maintain the virtual network connections, respondents showed that they prefer to stay in accommodations that had free internet or wi-fi access.

Table 4.7 presents the composite scores and the results of ANOVA for during-trip questions for the four age cohorts. All respondents stayed in contact with home using email. The use of SNS to stay in contact with home networks, however, was greater for the younger age cohorts, with the 18–24 cohort using SNS the most. The 40+ cohort was not likely to

**Table 4.6** Responses to during-trip questions

| *Item* | *Mean (n =176)* | *SD* |
|---|---|---|
| Use email to stay in contact with friends and relatives back home | 4.51 | 0.778 |
| While travelling, your email address and/or social networking profile is your only real address | 3.89 | 1.207 |
| Prefer to stay at hostels with free internet or wi-fi access | 3.76 | 1.220 |
| Added friends met during your trip to Facebook, Myspace, Bebo, etc. | 3.73 | 1.428 |
| Made future travel bookings online | 3.70 | 1.211 |
| Keep a journal | 3.65 | 1.256 |
| Use social networking site to stay in contact with friends and family back home | 3.61 | 1.485 |
| Changed travel plans during your trip after finding information online | 3.56 | 1.109 |
| Send postcards or letters, use calling cards, or call centers to stay in contact with friends and family back home | 3.49 | 1.251 |
| Used online backpacker forum to find information for the rest of your trip | 3.39 | 1.318 |
| Posted pictures online during your trip | 3.38 | 1.441 |
| Made a post to online forum hostel review site to share your experiences | 3.23 | 1.312 |
| Used a cell phone while travelling to stay in contact with friends and family back home | 2.95 | 1.515 |
| Maintain an online blog for friends, family, and other backpackers to read about your trip | 2.82 | 1.433 |
| Use a cell phone to stay in contact with fellow travellers or to contact accommodations, attractions, airlines, etc. while travelling | 2.79 | 1.449 |
| Stay in contact with work while travelling using email, cell phone, etc. | 2.45 | 1.413 |

**Table 4.7** Comparison of during-trip questions between age groups

| Travel behavior | 18–24 | 25–30 | 31–40 | 41 + | F value | Sig. |
|---|---|---|---|---|---|---|
| Use email to stay in contact with friends and relatives back home | 4.68[1] | 4.52[12] | 4.47[12] | 4.21[2] | 2.932 | 0.035 |
| Use social networking site to stay in contact with friends and family back home | 4.18[1] | 4.07[1] | 3.60[1] | 1.91[2] | 28.866 | 0.000 |
| Added friends met during your trip to Facebook, Myspace, Bebo, etc. | 4.32[1] | 4.04[1] | 3.80[1] | 2.09[2] | 28.729 | 0.000 |
| Posted pictures online during your trip | 3.71[1] | 3.59[1] | 3.27[12] | 2.56[2] | 5.641 | 0.001 |
| Used online backpacker forum to find information for the rest of your trip | 2.97[1] | 3.65[2] | 3.57[12] | 3.39[2] | 3.855 | 0.011 |
| Prefer to stay at hostels with free internet or wi-fi access | 4.00 | 3.89 | 3.43 | 3.38 | 2.924 | 0.035 |
| Made a post to online forum or hostel review site to share your experiences | 2.85 | 3.43 | 3.40 | 3.56 | 3.230 | 0.024 |
| While travelling, your email address and/or social networking profile is your only real address | 4.18[1] | 3.91[12] | 3.93[12] | 3.26[2] | 4.615 | 0.004 |

*Note:* Superscript numbers denote significant difference between means at $p < 0.05$. Tukey's HSD was used. Likert scale used (1 = low, 5 = high).

use SNS to stay in contact with friends and family. SNSs were also used to add friends met while travelling by all the cohorts except the 40+ group, topped by the 18–24 group. Online backpacker forums were used more often by the older age cohorts, 25–40+, to find information while travelling. Similarly, the sharing of information and experiences by posting to an online backpacker forums or hostel review sites was greatest for the 25–40+ cohorts. The posting of pictures to share experiences while travelling was greatest for the 18–24 cohort, and high for the two middle cohorts, but very low for the oldest. The perception that an email address or SNS profile is a backpackers only real address is greatest for the 18–24 cohort and decreases for each older cohort, which corresponds to the preference for hostels that offer free internet or wi-fi access, suggesting that many backpackers have a need to stay connected to their networks no matter where they are.

Backpackers are bringing more technology with them while they travel. Table 4.8 presents the items that backpackers most often bring along. Digital cameras are used by almost all backpackers in order to document their experiences. These pictures are not only for their own memories, but also to share with friends and family upon returning home, and to virtually share with the backpacker culture. As mentioned above, some backpackers even post pictures online in order for their friends, family and other travellers to virtually take part in their experiences. Backpackers are also starting to bring laptops and internet-enabled cell phones and PDAs with them while travelling. These devices allow them to use the expanding availability of wireless internet in backpacker hostels as well as in the other locations they frequent while travelling. Internet usage of backpackers while travelling

**Table 4.8** Technology on the road ($n = 168$)

| Technology | Frequency | % |
|---|---|---|
| Digital camera | 162 | 96.4 |
| Laptop | 35 | 20.8 |
| Video camera | 11 | 6.6 |
| Cell phone | 101 | 60.1 |
| IPod or other Mp3 player | 109 | 64.9 |
| PDA | 6 | 3.6 |

**Table 4.9** Internet usage while travelling

|  | *Frequency* | % |
|---|---|---|
| **How often do you log onto the internet while travelling?** | | |
| Several times a day | 12 | 7.1 |
| Once a day | 35 | 20.8 |
| Once every few days | 121 | 72.0 |
| **How long do you spend online each time you log on?** | | |
| Less than 30 minutes | 41 | 24.4 |
| 30 minutes to 1 hour | 104 | 61.9 |
| More than 1 hour | 23 | 13.7 |

is presented in Table 4.9. Most backpackers log on to the internet once every few days, and spend between 30 minutes and 1 hour online.

## Post-trip results

Composite scores for responses to the post-trip questions are presented in Table 4.10. Backpackers' post-trip virtual activities are concerned with enhancing and maintaining their virtual identity and virtual interacting with the backpacker culture. On returning home, many respondents uploaded their own photos and viewed other travellers' photos that they met during the trip. They also maintained contact via email and SNS with the friends they made. Online communities provided the arena for the sharing of experiences and information. The results suggest that through these shared experiences, backpackers are enticed to travel more. The virtual proximity that online communities allow like-minded people is also evident in the findings. Many respondents agreed strongly that online backpacker communities and forums allowed them the chance to have direct contact with the backpacker culture. This connection creates the sense of belonging and enables backpackers with large geographical distances between them to maintain close connections, allowing them to share their experiences, which many agreed is more fulfilling than sharing their experiences with their geographically close networks.

Table 4.11 presents the composite scores and the results of the ANOVA for the post-trip questions. The post-trip questions were concerned with the connection to the backpacker culture and the sharing of experiences.

**Table 4.10** Responses to post-trip questions ($n = 168$)

| Item | Mean | SD |
|---|---|---|
| Uploaded pictures to share with friends, family and fellow travellers | 4.24 | 1.010 |
| Viewed pictures/videos posted online by other travellers met during trip | 4.10 | 1.062 |
| Use email to stay in contact with friends and fellow travellers met during trip | 3.93 | 0.958 |
| Other backpacker's experiences shared online will entice you to travel more | 3.68 | 1.143 |
| Use SNS to stay in contact with friends and fellow travellers met during trip | 3.67 | 1.391 |
| Contributed to online backpacker forum | 3.57 | 1.311 |
| More fulfilling and/or easier to share experience with other travellers virtually than friends and family at home | 3.55 | 1.188 |
| Online travel communities provide direct access to the backpacker travel community | 3.54 | 1.077 |
| You stay connected to backpacker culture online while at home | 3.43 | 1.217 |
| You reviewed a hostel you stayed at on a hostel booking website | 3.01 | 1.325 |
| You joined a backpacker group on SNS after returning home | 2.95 | 1.422 |
| After kept maintain blog | 2.52 | 1.252 |

Email was used by all the age cohorts to stay in contact with travellers met on the road. The youngest cohort was more likely to use SNS to stay in contact with travellers met on the road, and the oldest cohort was not likely to use SNS. All four scores for uploading pictures to share with friends, family and fellow travellers were high, with the two youngest cohorts the most likely to post pictures online. Similarly, the two youngest cohorts were the most likely to view other travellers pictures online. Staying connected to the backpacker culture after the trip also occurred through participation in online communities. Backpacker forums were contributed to the most by the oldest cohort, and

**Table 4.11** Comparison of post-trip questions between age groups

| Travel behavior | 18–24 | 25–30 | 31–40 | 41 + | F value | Sig. |
|---|---|---|---|---|---|---|
| Use email to stay in contact with friends and fellow travellers met during trip | 3.86 | 4.09 | 3.96 | 3.81 | 0.696 | 0.556 |
| Use SNS to stay in contact with friends and fellow travellers met during trip | $4.20^1$ | $4.04^1$ | $3.70^1$ | $2.06^2$ | 27.164 | 0.000 |
| Uploaded pictures to share with friends, family and fellow travellers | $4.45^1$ | $4.47^1$ | $4.00^{12}$ | $3.69^2$ | 5.874 | 0.001 |
| Viewed pictures/videos posted online by other travellers met during trip | $4.41^1$ | $4.36^1$ | $3.85^{12}$ | $3.31^2$ | 10.532 | 0.000 |
| Contributed to online backpacker forum | $2.84^1$ | $3.91^2$ | $3.70^2$ | $4.44^2$ | 15.539 | 0.000 |
| You joined a backpacker group on SNS after returning home | 3.16 | 3.27 | 3.00 | 2.06 | 5.824 | 0.001 |
| You reviewed a hostel you stayed at on a hostel booking website | 2.86 | 3.22 | 3.26 | 2.78 | 1.304 | 0.275 |
| More fulfilling and/or easier to share experience with other travellers virtually than friends and family at home | 3.63 | 3.69 | 3.44 | 3.28 | 0.904 | 0.440 |
| Online travel communities provide direct access to the backpacker travel community | 3.44 | 3.73 | 3.63 | 3.38 | 0.988 | 0.400 |
| You stay connected to backpacker culture online while at home | 3.23 | 3.73 | 3.56 | 3.31 | 1.699 | 0.169 |
| Other backpacker's experiences shared online will entice you to travel more | 3.67 | 3.87 | 3.78 | 3.38 | 1.230 | 0.301 |

*Note:* Superscript numbers denote significant difference between means at $p < 0.05$. Tukey's HSD was used. Likert scale used (1 = low, 5 = high).

significantly less by the youngest cohort. The results also suggest that many people, with the exception of the oldest cohort, joined backpacker-specific groups within SNS. All the cohorts agreed to an extent with the statement that it is more fulfilling and/or easier to share experiences with other travellers virtually than with friends and family at home. Further, the virtual connection to the backpacker culture is supported by the responses of all four cohorts to the statements that 'online travel communities provide direct access to the backpacker travel community' and 'you stay connected to backpacker culture online while at home'. The sustained connection to the backpacker culture can entice backpackers to travel more as they experience other backpackers' experiences online.

## Discussion

Information is powerful, and is central to backpacking. In backpacker culture, the possession and ability to pass on insider information is a status builder among other backpackers. The recent emergence of online communities has allowed backpackers to centralize and virtualize both the social interactions and the transfer of information. Online forums like Lonely Planet's Thorn Tree Forum have created arenas in which the transfer of information can occur between travellers and non-travellers. The traditional transfer of information for backpackers was a result of fleeting face-to-face interactions while travelling on the backpacker trail and from other traditional sources, such as bulletin boards in hostels and guidebooks. While these sources are still used, and still play a large role in the maintenance of 'road status', the virtual online communities, as well as other online information sources, have become more popular sources of information. Indeed, online information enables greater mobility while on the road. Access to information and the ability to book travel on their own *prior to* departure allows backpackers that are temporally constrained, whether by a job, school or family, the ability to maximize their experiences during their trip. Further, independence *while* travelling is enhanced by the availability of just-in-time information, allowing for even greater flexibility in making travel decisions on the road.

The convergence of communications technology and social life has resulted in the centralization of people's multiple networks. SNSs, email, internet phones, instant messaging and other new technologies have allowed people to manage and stay in constant contact with their multiple overlapping networks. These networks can consist of friends, family, school, work, hobby groups and other communities. The virtual proximity of these networks allows for instant contact as well as simultaneous contact, no matter the geographical distance. Backpackers have embraced these technologies as well and many now maintain constant connection with their other networks while travelling. Email, SNSs, online communities, cell phones, laptops and wi-fi access all contribute to the close virtual proximity to personal networks. Moreover, the ability to instantly be in contact with home has contributed to the mainstreaming of the backpacker market. Parents of first-time backpackers can now feel more at ease, as can first-time backpackers, with the ability to be in instant contact with family members. Backpackers can stay in contact with work if they choose too, or in extreme cases they can maintain a constant nomadic status and work from the road, thus creating a mobile lifestyle. In these extreme cases, the person's email address or social networking profile becomes their 'true' address. The boundaries between home and away thus become blurred, and instead the distinction becomes between a person's virtual and physical identity.

In general society, the profusion and popularity of online communities, especially SNSs, allows regular people the same freedom that backpackers experience while travelling. SNSs and other forms of online identity, such as screen names, blogs, personal website and online photos, allow people to be seen online how they want to be seen. It provides a sense of control over their virtual personal identity that can be intoxicating. And backpackers have also virtualized their identities. While many of these identities are the same for multiple networks, as in the case of SNSs, many are particularly focused on enhancing social status and increasing social interaction within the backpacker culture, such as backpacker-specific personal blogs, websites and posts to online forums. Moreover, backpackers maintain a virtual connection to the backpacker culture after their trips. The sharing of information, pictures and experiences contributes to stabilization of the backpacker culture online.

This study sought to examine the virtualization of the backpacker culture that has resulted from the evolution of the culture's social structure. This evolution is a result of a combination of several factors, including an adaptation to the recent innovations of the internet and communications technologies, the overall mainstreaming of backpacking and the stability of the backpacker ideology. The results of the study support the argument that advances in the internet and communications technologies, particularly online communities, have indeed created a social structure that has allowed backpackers to maintain a constant connection to the backpacker culture.

Despite the mainstreaming of backpacker tourism, backpacker ideology (Welk, 2004) has remained remarkably stable over the last few decades. The findings of this study suggest that the virtualization of the backpacker culture has allowed backpackers to fulfill the actual practice of the five pillars of backpacker ideology to a greater extent. The ease of booking hostels and transportation online, especially through online booking agents, has allowed backpackers to find the best deal by comparing many options, both before departure and while on the road. The availability of just-in-time information has increased backpacker use of the internet for pre-trip internet searches, as well as supported increased independence and mobility while on the road. Individual backpacker's freedom and individual identities are enhanced through the ability to gain this information anywhere anytime, as well as to represent themselves how they want to be represented on SNSs and personal blogs. A large part of the appeal of backpacking is the freedom to be whoever one wants to be, and this ideal has also contributed to the adoption of SNS and blogs by many backpackers. Finally, the social interaction between backpackers and the opportunity to meet other people from around the world is arguably the most important backpacker ideal. The often fleeting and temporally constrained interactions that occur on the road can now be maintained through online communities and other forms of internet-based communications *post-trip*.

This study also incorporated an age-cohort analysis to examine differences between backpackers of different ages. This was conducted primarily because it could show how backpackers of different ages used the different types of communications technologies. Also, it can be assumed from the correlation between backpacker travel experience and age (Paris & Teye, 2009), that many of the older backpackers first 'became backpackers' when they were younger. The older cohort represents backpackers, who as younger travellers travelled during the 'hippie trail' period and have sustained the backpacker ideology. The virtual communities allow them now to reconnect to that culture through the internet. Some of this cohort are not likely to travel in the same manner as they did in their earlier backpacker experiences, as they probably have greater disposable income and a minimal need to maintain the 'badges of honor' related to 'roughing it' and travelling on a low budget. While they maintain some of the backpacker ideology and still have a personal connection to the backpacker culture, their actual connection on the road with backpackers may be minimal.

The older cohorts were more likely to take part in online backpacker forums, like Lonely Planet's Thorn Tree Forum. The 40+ cohort, in particular, use the forums for information, as well as contributing to the forums, suggesting that they maintain their connection to backpacker culture by passing on knowledge, gained through their previous experience, to future cohorts. This is similar to the role that guidebooks, and prior to that word-of-mouth, have played in the past, transferring knowledge from experienced older backpackers to younger less-experienced backpackers (Sorensen, 2003). Backpackers in the youngest cohort are not as likely to actively participate in online forums; however, they are very likely to use SNSs to maintain their connections to the backpacker culture. This connection via SNS results in the centralization of the younger backpackers' various networks into one place. SNS provide the platform for people to maintain a constant state of co-presence with all of their networks. Further connections to the backpacker culture are represented by the viewing and sharing of personal photos and personal blogs online. These, in a sense, play the role of maintaining the social status of the backpacker in the hierarchy of the backpacker culture. This is similar to the 'road status' that backpackers may seek to maintain based upon the level that the individual fulfills of the classic backpacker ideology.

## Conclusion

This study followed up on the agenda laid out in the concluding chapter of *Backpacker Tourism: Concepts and Profiles* (Ateljevic & Hannam, 2008). Through the contemporary mobilities lens, the convergence of communications technology and backpacker travel has been examined. The results of this study suggest that the convergence of the ideology of backpackers with the advances in communications technologies have created a social structure to support the mainstream backpacker culture of today. This study only took a partial look at this complex phenomenon. Backpackers were only targeted online; in order to strengthen the results, this study should be repeated through the administration of destination-based surveys at multiple backpacker destinations. Further, to complement the broadness of the quantitative results of this study, future qualitative studies could provide additional depth to the findings.

## Chapter 5
# Reconceptualising Lifestyle Travellers: Contemporary 'Drifters'

SCOTT COHEN

## Introduction

The question of whether there is a difference between the notions of 'traveller' and 'tourist' has been an ongoing area of debate in tourism research (Boorstin, 1964; Dann, 1999; Galani-Moutafi, 2000; Jacobsen, 2000; Welk, 2004), as well as being a vibrant discussion topic amongst travellers themselves (O'Reilly, 2005; Riley, 1988). Not only has it been reported that travellers self-define themselves as different than tourists (O'Reilly, 2005; Riley, 1988), but also within the leisure traveller population itself, commonly known as 'backpackers' (Loker-Murphy & Pearce, 1995), niches have developed that indicate that this group is not homogenous, but in fact multifaceted (Nash, 2001; Sørensen, 2003).

Sørensen (2003) has suggested that the heterogeneity of the backpacking market justifies further research into its specific sub-types. While research has been undertaken that investigates what may be termed the 'contemporary backpacker' (Noy, 2004; O'Reilly, 2006; Sørensen, 2003; Uriely *et al.*, 2002), Cohen (2004a) has noted that there is a dearth of research on 'contemporary drifters', those who seek to set themselves apart from mainstream backpackers, just as backpackers define themselves in opposition to tourists. Within the backpacker label, there exist a small proportion of individuals for which travel is not a cyclical break or transition to another life stage. For these individuals, leisure travel can serve as a way of life that they may pursue indefinitely. This chapter conceptualises these individuals as 'lifestyle travellers', a less pejorative term than 'drifter' (Cohen, 1972) or 'wanderer' (Vogt, 1976), which I have chosen to describe individuals that engage in long-term travel as a lifestyle.

The chapter begins with a critical review of the supposed traveller/ tourist divide and a consideration of the role of anti-tourism in constructing traveller identities. It then traces the development of the concept of the traveller over time, from Cohen's (1972) drifter, a term described in detail below as an idealised form of wanderer that has inspired modern backpacker idealism, through to the current state of the backpacking literature. This is done in order to provide a historical basis from which lifestyle travel has emerged that serves to situate lifestyle travel in relation to the theory on backpacker tourism. In finally delineating travel as a lifestyle, the use of lifestyle theory is employed to justify lifestyle travellers as an identifiable lifestyle group.

## The Traveller/Tourist Divide

Attempts at a distinction between traveller and tourist have regularly resurfaced in the academic literature and electronic discussions among tourism professionals (Dann, 1999). One method of exploring a supposed division has been etymologically. Fussell (1980) explained that 'travel' is derived from the word 'travail', which in turn was taken from the Latin 'tripalium', referring to a three-staked instrument of torture designed to rack the body. Thus, travel has been conceived as 'laborious or troublesome' and the traveller 'was an active man at work' (Boorstin, 1964: 84–85). By contrast, and appearing later chronologically, the word 'tourist' was derived from the Latin 'tornus', based on a Greek word for a tool that describes a circle (Boorstin, 1964). As such, Boorstin (1964) purported that the nature of travel changed with the decline of the traveller and the rise of the tourist, with the former working at something and the latter being a pleasure-seeker.

An association between travel and work can be linked to the Grand Tours of 17th- and 18th-century Europe, undertaken primarily by affluent young males (Loker-Murphy & Pearce, 1995). These tours were depicted as a form of education, a finishing school in which travel was intended to increase one's worldliness, social awareness and sophistication (Loker-Murphy & Pearce, 1995). As such, the Grand Tour was romanticised as being of educational value, rather than as a domain for seeking pleasure. Whether these tours were, in fact, also hedonistic, is beside the point for the moment, as the focus here is on how a privileged class composed mainly of well-off, white, European males (Galani-Moutafi, 2000) came to view themselves, at least outwardly, as elite travellers enhancing their education.

Further seeking to show how distance came to be perceived between a traveller and a tourist, Fussell (1980: 39) suggested that the Grand Tour pre-dated tourism, having pointed out that, 'before the development of tourism, travel was conceived to be like study, and its fruits were considered to be the adornment of the mind and the formation of judgment'. It comes as no surprise following this statement that the eventual development of mass tourism was viewed by many in the upper classes of Western society as signifying the end of the possibility of 'real' travel (Boorstin, 1964). In his essay on the *Lost Art of Travel*, Boorstin (1964) lamented the growth of mass tourism, which was largely attributed to technological developments such as the railway. Whereas transportation had previously been quite laborious, the industrial mass production of long-distance transportation, via railways and ocean steamers, made travel purportedly more pleasurable, cheaper and accessible (Boorstin, 1964). As Boorstin (1964: 86) complained, 'thus foreign travel ceased to be an activity – an experience, an undertaking – and instead became a commodity'.

Boorstin's (1964) lamentation of the decline of the traveller and the rise of the tourist helped to produce a dichotomy in the tourism literature within which the disappearing traveller was awarded the nostalgic prestige of earlier explorers and the proliferating tourist experience was pejoratively described as contrived, diluted and prefabricated. Thus, for some, the traveller can be positively valued while the tourist can be perceived negatively (O'Reilly, 2005). Fussell (1980) has pointed out that the negative connotations ascribed to the word 'tourist' are even perpetuated within the tourism industry itself, as travel agencies elect to be called *travel* agencies rather than *tourism* agencies.

Moreover, Dann (1999) has recognised that many literary works have managed a traveller versus tourist distinction that parallels that of the sacred versus profane, and that this supposed distinction has allowed for broadly contrasting analogies of types of tourism as either a sacred journey (Graburn, 1977) or as a game or form of play (Lett, 1983). Within a traveller/tourist dichotomy, wherein travel is depicted as more 'real' or superior to tourism, it has become clear that the idea of the traveller is heavily based on anti-tourism and tourist angst (Dann, 1999).

## Anti-tourism in the Construction of Traveller Identities

Dann (1999) suggested that the idea of travel is simply anti-tourism. Richards and Wilson (2004a: 49) pointed out that in forming selves, 'it is often easier to say clearly what one is not than what one is'. It has been noted that within the context of ethnicities, differentiation is often based against the community that lives closest to one's own (Welk, 2004). Welk (2004) also suggested that while a distinction can normally be made in territorial terms, it is also possible to differentiate a group based on a perceived symbolic basis. In the case of travellers, the closest perceived group in terms of form is mainstream tourists (Welk, 2004 and Richards & Wilson, 2004a made this comparison except having used the term 'backpacker' rather than traveller). In this light, Bruner (1991: 247) alleged that a distinction between a tourist and a traveller can be reduced to 'a Western myth of identity'. Consequently, a traveller identity is largely constructed through its opposition to the tourist role, or in other words, in opposition to what it hopes not to be.

Accordingly, while travellers may view the tourist experience as clichéd and devoid of spontaneity, by contrast, the traveller's experience is (re)produced as risky, exciting and imbued with freedom (Dann, 1999; Fussell, 1980). Welk (2004) further urged that the issue is best clarified by looking back in history to cultural origins in the hippie movement. The counterculture that Cohen (1972) based his conceptualisation of the 'drifter' on was a reaction against a 'conformist' parent generation, and as such, any divide between travellers and tourists has roots in both a class and generational conflict (Welk, 2004). However, Fussell (1980: 49) critically surmised that the anti-tourist conviction that one is a traveller instead of a tourist is 'both a symptom and a cause of what the British journalist Alan Brien has designated *tourist angst*, defined as "a gnawing suspicion that after all...you are still a tourist like every other tourist"'. For Fussell (1980), tourist angst is a class signal, extending back as described above to the Grand Tour, in which the anti-tourist, or traveller, is deluding herself/himself. As Fussell (1980: 49) continued 'We are all tourists now, and there is no escape'.

Thus, the question emerges as to whether travellers are distinct from tourists or, in fact, just tourists themselves. This can be addressed by both looking at what travellers say about themselves and deeper into the differences that the literature has suggested. With both of these methods, it is useful to clarify in advance between form and type-related attributes (Uriely *et al.*, 2002). Form-related attributes refer to the arrangements and practices through which an individual constructs a journey, such as style of accommodation or length of trip. On the other hand, type-related attributes are of a psychological nature, for instance, the meaning one assigns to a trip or motivations for travel (Uriely *et al.*, 2002).

In terms of what travellers say about themselves, or self-definition, Welk (2004) claimed that travellers do not necessarily perceive them-selves as 'better' tourists, but reject the tourist label altogether in exchange for the term traveller. Likewise, O'Reilly (2005) reported that traveller is the term preferred by most backpackers. In White and White's

(2004: 202) study of mid-life and older long-term travellers in the Australian Outback, the terms *traveller* and *travel* were used in place of *tourist* and *touring* because the authors felt that the former terms 'more accurately capture the meaning of these journeys for those undertaking them'. This was also evidenced in Richards and Wilson's (2003) study in which over half of a 2300 person sample identified themselves as travellers, around one third as backpackers and only one fifth as tourists. Particularly notable within Richard and Wilson's (2003) study was that younger persons were much more willing to accept the backpacker label than older persons.

In Davidson's (2005: 35) study of travellers in India, the majority of respondents were unhappy to describe their travels as 'backpacking', as the term 'now conjures up images of young, privileged gap-year students'. In this sense, most of Davidson's (2005: 35) interviewees would have reportedly perceived the term 'backpacker' as 'an insult to their status as travellers'. Moreover, in Riley's (1988) study of long-term travellers, 100% of the interviewees rejected the tourist label, having justified themselves as travellers based on form differences of available time and money. Long-term travellers were conceptualised as rich in time, but financially constrained. Dann (1999) distinguished between travellers and tourists by duration as well – having argued that tourists generally have less time at their disposal.

On the other hand, O'Reilly (2005) found that some travellers admitted there is not a real form difference between tourists and travellers, but did claim that there are significant variances in their respective approaches to travel. Her interviewees cited the difference in approach to the individual's openness to the experience as a 'journey of self' (type) rather than upon money or style of accommodation (form). In further highlighting that such type differences might exist between travellers and tourists, in Maoz's (2004: 114) study of Israeli travellers in India, interviewees reported they were on a 'serious and profound inward journey' with a desire to 'find themselves'; a quite stark contrast to Boorstin's (1964) contrived tourist experience.

Thus, in regard to type-attributes, for travellers the distinction is made in that it is the journey rather than the destination that is important, as travel may be considered as an 'inward voyage' or 'state of mind' that embodies feelings of independence and freedom (Galani-Moutafi, 2000; O'Reilly, 2005). Galani-Moutafi (2000: 205) described how an inward voyage may mirror the external physical journey wherein 'a movement through geographical space is transformed into an analogue for the process of introspection'. By contrast, being a mass tourist purportedly may lack this transformative power (O'Reilly, 2005). With this in mind, Welk (2004: 90) re-emphasised the importance of time in journeying as a state of mind by having alleged that 'the difference between the tourist and the traveller can be seen in that tourism is a temporary state of existence, while travelling is a permanent one'. However, to draw a line between travel and tourism based solely on the presence or not of a sense of inward voyage lays the distinction between traveller and tourist within subjectivism.

Even though academia has not come to agreement as to whether travellers and tourists do differ, it is clear enough that both form and type differences have been suggested in the literature. While these differences surely do not hold fast across all individuals nor represent entirely distinct categories, it is likely that they do represent some broad trends that allow for the notion of the traveller to be teased out as a sub-type of the tourist. However, even if the form and type criteria for a traveller are not taken to differ significantly from the mass tourist (see Sharpley, 2003), it is hard to overlook the emic perspective of the traveller, who has reportedly identified with the self-definition of traveller rather than tourist (O'Reilly, 2005; Riley, 1988; Welk, 2004). Hence, anti-tourism allows for the construction of a distinct identity even if the reality may be taken as a contradiction (Welk, 2004). Jamieson (1996) suggested that a perception of identity or status apart from other tourists is what, in fact, sets travellers apart from other tourists.

Nash (2001) has moved past the question of a traveller/tourist divide and suggested that travellers themselves are not even a homogenous group, and as such, it is important to specify which type of traveller one is talking about. In a similar vein, this chapter now does the same. In order to situate the concept of a lifestyle traveller in relation to the broader contemporary context of the 'backpacker', the focus now turns to how the tourism literature has historically framed the traveller as the antecedent to the backpacker, beginning with Cohen's (1972) conceptualisation of the 'drifter'.

## From Drifter to Backpacker and Beyond

Some of the academic (Cohen, 1982a, 2003; O'Reilly, 2006) and popular literature (Garland, 1997; Sutcliffe, 1998) concerning travellers has highlighted a supposed institutionalisation of the backpacking phenomenon, comparing contemporary backpackers with conventional mass tourists, with the academic side normally tracing this development from its theoretical source in Cohen's (1972) drifter ideal. To a degree, these works have lamented the loss of the drifter ideal and have helped to increasingly compartmentalise travellers as mainstream backpackers, resembling mass tourists, to the point where the term 'backpacker' has nearly replaced the word 'traveller' in certain regions (Loker-Murphy & Pearce, 1995). However, the literature has at times recognised that the backpacker market is heterogeneous (Nash, 2001; Sørensen, 2003; Uriely *et al.*, 2002), which has opened the door to the study of its sub-types.

This chapter now turns to the academic development of the drifter role, in order to examine how it has inspired the mythology of contemporary backpacking ideals. Rather than laying the drifter ideal to rest and homogenising all travellers under the broader rubric of the institutionalised backpacker, however, it is suggested that some travellers repeatedly exceed the temporal boundaries that have traditionally situated backpacking as transitory, a rite-of-passage and/or a liminoid experience (Lett, 1983; Turner, 1977). Hence, it is argued that some travellers have established travel as their 'normal' way of life or lifestyle, rather than a break from it.

In outlining the historical origins of backpacker tourism, some researchers have pointed to the wider history of tourism and again located the starting point for backpacker theory within the 17th- and 18th-century European Grand Tour (Loker-Murphy & Pearce, 1995). Alternatively, others have begun a historical tracing of backpacking theory with Cohen's (1972) conceptualisation of the drifter as the archetypal backpacker (Richards & Wilson, 2004b). These divergent starting points offer quite different connotations, because while the Grand Tour is associated with education and supposed sophistication, drifting often carries derogatory connotations (Riley, 1988).

A careful analysis of the literature does locate a degree of historical importance to the Grand Tour, but further questions how what was once largely the reserve of affluent upper-class youth came to bear deviant associations in the form of the drifter. Part of the answer may be found by examining another important pre-cursor to the backpacker, which is oft overlooked in backpacker studies, the 19th-century 'tramp', a working-class, young, adult male, who through vocational membership, followed a circuit of small-town craft society inns that supplied accommodation and work (Adler, 1985). Distinguishable from the Grand Tours of affluent upper-class youth, the tramping system was borne out of economic necessity among the working class. At the level of motivation, however, both the Grand Tour and the tramping system supplied a ritual separation from family to foment the transition into adulthood, as well as providing the opportunity for adventure and education (Adler, 1985).

But increased urbanisation and industrialisation around the beginning of WWI spelled the decline of organised craft associations (Alder, 1985). Many young, working-class males continued the tramping form, but as they were no longer legitimised by the craft societies, their mobility came to be viewed as a social problem, almost a type of vagrancy (Adler, 1985). Adler (1985) suggested that the literature on youth travel shifted to represent this new image of travelling youth as a form of juvenile delinquency. Hence, while tourism among the middle classes grew with the establishment of mass produced long-distance transportation in the 20th century (Boorstin, 1964) and travel continued to play a role in the lives of young people, for the latter it changed in that their travels began to be perceived by society as hedonistic and even anarchistic – a form of escapism (Loker-Murphy & Pearce, 1995). This was also largely due to an increasing association with the 'hippie counterculture' of the 1960s (O'Reilly, 2006).

The 1960s counterculture largely stemmed from a generational values conflict between dissatisfied youth and an allegedly conformist parent generation (Welk, 2004), and as such, was 'both a symptom and expression of broader alienative forces' (Cohen, 1973: 94). A large section of alienated youth in developed Western countries in the 1960s were mainly articulating a political statement against the growing cultural and political homogeneity of the period, with one primary area of expression taking place through opposition to the Vietnam War (Ateljevic & Doorne, 2004). Seemingly running counter to the values of mainstream society, the counterculture became associated with rebellion, drug usage and anarchic values (Ateljevic & Doorne, 2004), but to the youth themselves, their movement represented a mission, a chance to revolutionise their home societies (Welk, 2004).

One distinctive means through which countercultural adherents expressed themselves was through travel, as travelling fitted 'admirably the style of life and the aspirations of the members of the "counter-culture"' (Cohen, 1973: 93). Even popular literary works of the time, such as Kerouac's (1957) *On the Road* and Hesse's (1964) *The Journey to the East*, helped to link countercultural values to mobility. In addition to travel in Europe and North America, it became increasingly common in the late 1960s and early 1970s for Western youth seeking what they perceived as a more meaningful or authentic cultural existence (Ateljevic & Doorne, 2004) to follow what became known as the 'hippie overland trail' from Europe to India and Nepal (MacLean, 2006; Tomory, 1996).

This overland route, followed by hundreds of thousands of young Westerners, began in Istanbul and crossed through Iran, Afghanistan,

Pakistan and India, often finishing in Nepal (MacLean, 2006). India and Nepal signified the end of the road for these travellers, as the counter-culture had idealised the Indian subcontinent as the pinnacle of 'alternative lifestyle' destinations (Tomory, 1996), and popular discourse about the subcontinent had promised adventure, an 'earthly utopia' and even enlightenment (MacLean, 2006). Indeed, with many disaffected countercultural youth seemingly in search of a utopia, More's (1516) *Utopia*, a literary piece that described a socialised utopian island nation and influenced the revival of hippie communes during the counter-cultural period (Baumeister, 1986), appears to have influenced these idealistic wanderers. Interestingly, it seems that for many young travellers, while the overland route had spiritual connotations that theoretically support tourism as a sacred journey or secular pilgrimage (Graburn, 1977), the route also, especially to its moral critics, was viewed as hedonistic.

As the overland trail from Europe to Asia grew in popularity with primarily young travellers, and alienated Western youth increasingly 'drifted' through other parts of the world, both popular literature and academia picked up on the phenomenon. Michener's (1971) *The Drifters* served as a fictional account of the travels of several alienated youth during the period and Cohen (1972) conceptualised the drifter type within tourism studies as part of a broader endeavour to establish a tourist typology. Cohen's (1972) seminal work on a sociology of tourism spawned not only several other of his own publications that expanded on the notion of the drifter, but also established the drifter as an archetypal or idealised traveller, currents of which still run through modern-day backpacker theory and ideology. As Cohen (2004a: 44) suggested: 'If the model for the drifter was the tramp, the drifter is the model for the backpacker'.

## The Drifter Ideal

Cohen's (1972) discussion of the drifter included his conceptualisation of the 'explorer', with both these types described as 'non-institutionalised' tourist roles that were loosely linked to the tourist establishment. A drifter was originally depicted by Cohen as:

> This type of tourist ventures furthest away from the beaten track and from the accustomed ways of life of his home country. He shuns any kind of connection with the tourist establishment, and considers the ordinary tourist experience phoney. He tries to live the way the people he visits lives, and to share their shelter, foods, and habits,

keeping only the most basic and essential elements of his old customs. The drifter has no fixed itinerary or timetable and no well-defined goals of travel. (Cohen, 1972: 168)

In contrast to the drifter, whom Cohen compared to the 'wanderer' of previous times (although without explanation as to who constituted the latter), Cohen (1972) described the explorer as the traveller of former years, who does not, however, try to identify with the 'natives' and become one of them during her/his stay.

The explorer arranges his trip alone; he tries to get off the beaten track as much as possible, but he nevertheless looks for comfortable accommodations and reliable means of transportation. Though novelty dominates, the tourist does not immerse himself completely in his host society, but retains some of the basic routines and comforts of his native way of life. (Cohen, 1972: 168)

Although Cohen (1972) equated the explorer with the traveller of former years, both his future research (Cohen, 1973, 1982a, 2004a) and much of the other research that traces the 'evolution' of backpacking (Maoz, 2004; O'Reilly, 2006; Richards & Wilson, 2004b; Welk, 2004) has instead centred on the notion of the drifter as the primary precursor to the contemporary backpacker. This was largely influenced by Cohen's (1973: 90) conclusion that drifting had moved in a short space of time from a minor phenomenon into 'one of the prevalent trends of contemporary tourism', as the most popular and widespread form of travel for the Western younger generation.

Cohen (1973) supported that the popularity of drifter travel was due to its strong connection with the counterculture. He suggested that the mobility of drifting aided in the loosening of ties and obligations, the abandonment of accepted norms and the search for sensual experiences (Cohen, 1973). Not surprisingly, as with the counterculture in general, 'drifting', a derogatory term in the eyes of 'mainstream' society, not unlike tramping, came to be associated with deviancy, as a marginal and unusual activity undertaken by society's 'dropouts' (O'Reilly, 2005, 2006).

In a further effort to typologise tourists, Cohen (1979) proposed that 'serious' drifters could be compartmentalised under what he referred to as 'experimental mode', which characterised individuals pre-disposed to trying out alternative ways of life as part of a quest for meaning. These experimental travellers were purported to be in 'search of himself' as part of a trial and error process (Cohen, 1979: 189). However, Cohen

(1982a: 221) lamented that most young travellers do not even qualify as part-time drifters, and even at this early stage in the drifter literature, he suggested that most youth travel 'in a conventional style characteristic of the institutionalized mass tourist', a statement that, as is discussed later, seeded much of the later backpacker literature that emphasised the homogenisation of the backpacker.

## Attempts at a Less Pejorative Term than 'Drifter'

Apparently, in an effort to defuse the hedonistic connotations of the drifter, Vogt (1976) and Riley (1988) attempted to conceptualise independent travel under different terminology. Vogt (1976) claimed that the primary motivation of wandering youth, as opposed to being aimless, was personal growth and, accordingly, called his young travellers 'wanderers'. Akin to Cohen (1979), Vogt characterised this type of travel as exploratory, offering ways to learn about the world and self and undertaken mostly by middle-class students on a moratorium from study. Vogt's (1976) wanderer concept seems to have been an attempt to realign travel with the educational value of the Grand Tour (O'Reilly, 2006), and to distance it from the negative associations of tramping and drifting.

In contrast to both Vogt (1976) and Cohen (1972), Riley (1988) protested the wanderer concept as being limited to relatively short-term young travellers, mainly composed of students, and also suggested that the derogatory drifter label had been misleading in connoting deviant behaviour. Riley's (1988) argument was based on a judgement that the demographics and motivations of the population engaging in independent travel had shifted in a generation to include individuals that could not be characterised as youth dropouts or just students on a break. In comparison, these travelling individuals were supposedly at a juncture in life, as opposed to generally aimless, hailed from middle-class backgrounds, and were, on average, older than the earlier travellers (Riley, 1988). Hence, unlike Cohen (1972) and Vogt (1976), who focused on young travellers, Riley (1988) found most of her travellers to be in their late 20s and early 30s, and with one of her interviewees being 60 years old, she placed no age restrictions on long-term travel.

Riley (1988) managed a conceptual distinction between travellers and shorter-term tourists, using Graburn's (1983) division of modern tourism into two modes. The first was annual vacations that 'mark the progress of cyclical time' and the second was 'rite-of-passage' tourism, which was described as taking place at the junction of major changes in life relating to, for instance, adulthood, career and/or relationships (Graburn, 1983: 12). Riley (1988) suggested the phrase 'long-term budget traveller' as a less pejorative and more accurate way of depicting individuals who engaged in rite-of-passage tourism. 'Long-term budget traveller' was also intended to reflect that the extended length of the individual's travel required most people to operate on a budget. As such, it was assumed that the tourist had limited time at her/his disposal, whereas the traveller was abundant in time, but was usually constrained by finances, and thus, self-imposed a budget in order to extend the travel period 'beyond that of a cyclical holiday' (Riley, 1988: 317).

In summary, Cohen's (1972) drifter, Vogt's (1976) wanderer and Riley's (1988) long-term budget traveller were each attempts to conceptualise travellers as distinguishable from tourists while representing what was perceived as the changing nature of independent travel. Although Vogt (1976) and Riley (1988) called for a term with less derogatory connotations than 'drifter', their respective suggestions of 'wanderer' and 'long-term budget traveller' were not embraced within academia, and it was not until Loker-Murphy and Pearce's (1995) introduction of the succinct label 'backpacker' into the academic literature that a less pejorative and more widely accepted term surfaced.

## The Growth of Backpacking and Backpacker Research

As long-term travel grew in popularity, the term 'backpacker' took hold from the late 1990s (O'Reilly, 2006) as a means of describing predominantly young, budget tourists on extended holiday (Loker-Murphy & Pearce, 1995). More structured than drifter travel, but purportedly different from mass tourism, backpackers supposedly displayed a preference for budget accommodation, an independently organised and flexible schedule, longer holidays, an emphasis on meeting other travellers and locals, and a penchant for informal recreation activities (Loker-Murphy & Pearce, 1995). More recently, Maoz (2007: 123) has described backpackers as 'self-organized pleasure tourists on a prolonged multiple-destination journey with a flexible itinerary', many of whom are on a transitory leave from relative affluence.

Historically, backpackers have been mainly of Western origin with the majority coming from Western Europe, Australia, New Zealand and North America (Maoz, 2007). However, with a significant population of Israeli backpackers, and a steady growth in Asian backpackers, particularly from Japan, the backpacker literature's conceptualisations have been increasingly criticised as being too Western oriented (Maoz, 2007; Teo & Leong, 2006). On the whole, growth in the backpacker industry has been substantial 'over the past 30 years, progressing from a marginal activity of a handful of "drifters" to a major global industry' (Richards & Wilson, 2004b: 10) that is now viewed by many as an accepted rite-of- passage for young people (O'Reilly, 2006).

Along with the growing popularity of backpacking in recent years, academic interest in backpacking has also grown (Hannam & Ateljevic, 2007a). It has been noted that backpacker research has generally been divided between sociological and market-based approaches (Richards & Wilson, 2004a, 2004b; Wilson & Richards, 2008a, 2008b). A number of studies have looked at identity-related issues in the context of back-packing, including themes of risk taking, adventure, anti-tourism and narratives of self-change (Anderskov, 2002; Desforgesm, 1998, 2000; Elsrud, 2001; Noy, 2004; O'Reilly, 2005; Welk, 2004). Additional research, amongst others, has delved into backpacking as a means of escape (Ateljevic & Doorne, 2000), form and type-related differences among backpackers (Uriely *et al.*, 2002), the culture of backpacking (Muzaini, 2006; Sørensen, 2003; Westerhausen, 2002), the roles of backpacker enclaves (Ateljevic & Doorne, 2005; Wilson & Richards, 2008a, 2008b), the specifics of Israeli backpackers (Haviv, 2005; Maoz, 2004, 2007; Noy & Cohen, 2005; Shulman *et al.*, 2006), social interactions of backpackers (Murphy, 2001), the influence of travel writing upon backpackers (Richards & Wilson, 2004a) and a postcolonial analysis of backpacking (Teo & Leong, 2006). Additionally, there have been a number of market-based and development studies focusing on backpackers (see, for instance, Richards & King, 2003; Richards & Wilson, 2003; Scheyvens, 2002; Westerhausen & Macbeth, 2003).

One other related theoretical point within backpacker research, as noted earlier in this section and expanded upon below, is that some of the literature has suggested an institutionalisation or 'mainstreaming' of the backpacking phenomenon (Cohen, 2004a; Hannam & Ateljevic, 2007b; O'Reilly 2006). This can largely be attributed to the commodification of the backpacker market, as backpackers are increasingly seen to follow the same distinctive trails, use the same guidebooks and gather in estab-lished backpacker enclaves, within which the backpacker 'scene is (re)produced' (Sørensen, 2003; Westerhausen & Macbeth, 2003; Wilson & Richards, 2008a: 188).

## The Institutionalisation of Backpacking

Many of the paths that backpackers follow are well-trodden in large numbers (Cohen, 2004a), and as such, a parallel travel system to mass tourism has developed that caters mainly to the backpacker market (Loker-Murphy & Pearce, 1995). Consequently, contemporary backpackers utilising these facilities have not necessarily had to develop the skills and invest the effort in their trips that were attributed to the earlier drifters (Cohen, 2004a). Cohen (2004a) has recognised that the figure of the drifter was, and still is (O'Reilly, 2006), an ideal towards which many youth are attracted, but very few have attained. As O'Reilly (2006: 1005) suggested 'the ideal typical backpacker of today closely resembles the model set down by the hippie travellers of the 60s and 70s'.

The difficulty of living up to the drifter ideal is explained by Cohen's (2004a: 45) admission that 'drifting, as I have conceived it, appears to take much more competence, resourcefulness, endurance and fortitude, as well as an ability to plan one's moves, even if they are subject to alteration, than I had originally surmised'. As mentioned previously, in contrast to the drifter concept, Cohen (1982a: 221) surmised that most young tourists, not then yet referred to as backpackers, travel in the style of 'the institutionalized mass tourist'. In view of that, Cohen (2004a) and O'Reilly (2006) have more recently decried the 'mainstreaming' of backpacking, as it now frequently carries the same stigma of institutionalisation associated with mass tourism, especially along the more popular backpacker routes of Australia and Southeast Asia (O'Reilly, 2006). However, as discussed in the following section, despite efforts to homogenise backpackers with mass tourists, there has been a growing body of literature that has attempted to highlight heterogeneity within the backpacker label (see Ateljevic & Doorne, 2000; Nash, 2001; Sørensen, 2003; Uriely *et al.*, 2002).

Similar to the tourist angst that travellers reportedly harbour, there appears to exist within backpacking culture angst towards institutionalised backpackers (Richards & Wilson, 2004a; Welk, 2004). This comes as no surprise considering that the backpacking ideal of the drifter is positioned in opposition to institutionalisation. Welk (2004) argued that the more involved travellers no longer strive to distinguish themselves from tourists, but instead from mainstream backpackers, especially the stereotype of the young party backpacker. Indeed, Wilson and Richards (2008a: 188) commented that this type of backpacker angst 'is prevalent among older, more experienced independent travellers lamenting the loss of their pioneering travelstyles due to the changing nature not only of tourism but also of backpacking'. In other words, as Welk (2004: 89) observed, for some travellers, 'anti-tourism has given way to anti-backpacking as their main category of distinction'.

As evidenced in Loker-Murphy and Pearce's (1995) recentring of the traveller discussion around younger travellers, travellers of all ages are often herded under the backpacker label. This continues despite Riley (1988) having urged that older travellers have also engaged in long-term travel as a rite-of-passage. Moreover, Westerhausen (2002) pointed out an increasing diversity of ages among backpackers as its appeal as a mode of tourism widened. Fittingly, Richards and Wilson (2004c: 65) observed that 'the backpacker as a clearly defined species of tourist is disappearing, just at the moment of its discovery'.

Ateljevic and Doorne (2000: 131) noted that studies focusing on the backpacker market tended 'to treat these travellers as a homogenous consumer group'. Ateljevic and Doorne (2000) instead suggested that there is heterogeneity under the backpacker umbrella term. Moreover, Uriely *et al.* (2002: 536) put forward that backpackers could be further classified into sub-types that might reveal similar groupings 'in terms of their motivations and the meanings they derive from travel'. Helping to determine whether different sub-types existed within the backpacker label, Uriely *et al.* (2002) conducted research into the backpacking population in terms of form and type-related attributes. While they concluded that backpackers share a common identity based on form-related practices (length of time on road, budget, form of transport), it was also found that backpackers are heterogeneous in type, displaying varying motivations and attitudes (Uriely *et al.*, 2002). These findings were supported by Sørensen (2003: 848) who has suggested that while backpacker facilities have become homogenised, its users seemed 'more composite and multifaceted than ever'. As such, Sørensen (2003) called for further research into specific sub-types of the backpacker market.

Richards and Wilson (2004a) have suggested that rather than seeing backpackers as part of a general tourist typology, it might be more fruitful to consider backpackers along a continuum of ideologies of its own. In this vein, Cohen (2004a) has noted that systematic research has not been undertaken into those travellers most closely reflecting the drifter ideal, who may seek to distinguish themselves from contemporary backpackers. Indeed, Uriely *et al.* (2002) claimed that only a minority of backpackers are travelling in Cohen's (1979) experimental mode, with many instead seeking the diversionary and recreational experiences of mass tourists. However, notably, among their (2002) interviewees were 'serial' or repeat backpackers who had often started their travels as recreational tourists and later switched into the experimental type. As Cohen (1979) has linked the experimental type with the more serious of the drifter segment, it is possible that Uriely *et al.*'s (2002) serial or repeat backpackers may be linked to Cohen's (2004a) notion of the 'contemporary drifter', at least in a temporal sense. This temporal component informs the next part of this chapter, which looks at the notion of travel as a way of life.

## Travel as a Way of Life

Generally, backpackers perceive their travels as a time out from their normal life-path (Elsrud, 2001) and/or as a self-imposed rite-of-passage (Graburn, 1983; Maoz, 2007), whether it occurs at the juncture between school and university, university and a career, or between careers (Cohen, 2004a). Maoz (2007) also suggested that some backpackers have experienced what may be described as 'life crises' prior to their journeys. However, while drifters of the 1960s and 1970s were described as 'alienated individuals roaming the world alone' in reaction to a perceived value conflict with Western societies (Cohen, 2004a: 44), estrangement from one's home society is purportedly a less central theme to most modern backpackers (Maoz, 2004). Correspondingly, the majority of backpackers expect to rejoin the workforce in their home society (Riley, 1988) and re-engage with the lifestyle they had left at home (Sørensen, 2003; Westerhausen, 2002), as few view travel as a feasible indefinite alternative to a 'normal' career (Cohen, 2004a). Even Cohen's (1972: 176) original drifter was characterised as eventually settling down to an 'orderly middle-class career' after a period of drifting.

However, in Riley's (1988) study of long-term travellers, a small number of individuals were identified that did treat travel like a career. Cohen (1979) also observed that some extreme cases, the more serious of the drifters, extended the search for meaning through travel into a way of life, essentially becoming 'eternal seekers'. Accordingly, Noy and Cohen (2005: 3) suggested that, for some, backpacking can cease to be a part of a transitional phase in life and can instead extend to 'a way of life in itself'. This is further substantiated by Giddens' (1991) assertion that as identity has become less staked out in post-traditional societies in late modernity, rites of passage as lifespan markers have become less relevant.

Likewise, Westerhausen (2002: 154) noted the phenomenon of a growing number of individuals travelling into and beyond their 30s, suggesting that what was once largely the domain of youth culture, 'now represents a lifestyle alternative for those at least temporarily unencumbered by family and professional responsibilities'. Westerhausen (2002: 146) summarised this type of dedicated backpacker best in stating that 'for a sizable minority, being on the road becomes a preferred way of life to which they will return whenever the opportunity presents itself'. Moreover, Welk (2004: 90) identified that for some backpackers 'the journey is not designed to be an interruption from normality, it is normality; and it is not supposed to serve any goals beyond travelling itself'. Welk (2004) used this theme to try to distinguish between backpackers and travellers, having claimed that a backpacker becomes a traveller when travel becomes a way of life.

Hence, it is suggested that some individuals do not fit within the cyclical and/or temporal boundaries that have traditionally circumscribed the annual holiday and also situated backpacking as a rite-of-passage that is a transitory or liminoid experience (Lett, 1983; Turner, 1977). As such, some repeat or serial backpackers (Uriely *et al.*, 2002) have inverted the traditional form of the cyclical holiday as a time-out from routine, and rather than treating tourism as a break, they have instead established travelling as a 'normal' way of life that they may pursue indefinitely. In this light, Graburn's (1983) two modes of tourism – cyclical annual holidays and rite-of-passage tourism, falls short.

Noy and Cohen (2005) have suggested that 'lifelong wanderers' are difficult to locate as they often try to avoid tourist facilities, and as such, have rarely been the subject of research. An individual who repeatedly returns to long-term travel and considers travel to be her/his way of life, can aptly, and less pejoratively than 'drifter' or 'wanderer', be termed a 'lifestyle traveller'. Lifestyle travellers reflect Cohen's (2004a) notion of 'contemporary drifters' and Noy and Cohen's (2005) 'lifelong wanderers'. As for the term 'traveller', as opposed to 'backpacker' or 'tourist', traveller is the self-defined label of most backpackers, especially older ones (Richards & Wilson, 2003), as unlike the term 'backpacker', it does not primarily connote transitional youth. Instead, 'traveller', as an identity perceived as set apart from a more temporally constrained 'tourist', emphasises a journey over time (Welk, 2004), with an 'assertion that travelling is a lifestyle or a "state of mind"' (O'Reilly, 2005: 158). In order to conceptualise travel as a type of lifestyle, and in doing so, further delineate lifestyle travellers as an identifiable sub-type of backpackers, it is now useful to turn to the literature on lifestyle theory.

## Travel in the Context of Lifestyle Theory

The origin of the term 'lifestyle' can be located as far back as Max Weber, but Wrong (1990: 24) has noted that within everyday usage, the word 'spread like wildfire at the height of the student protest movements of the late 1960s'. As sociology students played a strong role in the countercultural protests, it was partly their familiarity with Weber on 'style of life', which informed their movement for a break from the past and the adoption of 'alternative lifestyles' (Wrong, 1990). Within the academic literature, Veal (1993) recognised that a consensus had not been established on the meaning of the term lifestyle, as over 30 varying definitions had been offered. After a review of lifestyle theories within the context of its Weberian usage, as well as 'sub-cultural, psychological, market research and psychographics, leisure/tourism styles, spatial, socialist lifestyles, consumer culture, gender, and miscellaneous approaches', Veal (1993: 233) later suggested that the concept of lifestyle could be more clearly defined as 'the distinctive pattern of personal and social behaviour characteristic of an individual or a group' (Veal, 1993: 247).

It has also been suggested that most (Western) people seek 'coherence' in their lives, without necessarily finding it, and that many individuals are engaged in a 'life task' of establishing a set of activities or behaviours that 'make sense' to themselves (Veal, 1993). Furthermore, Giddens (1991) theorised that the concept of lifestyle has become progressively more important in modern social life as tradition has continued to lose hold and the increasing affects of globalisation have forced many individuals to negotiate a larger variety of life options. The breakdown of traditional roles, which previously contributed to a more secure sense of self, has made in turn, for some, lifestyle choice critical in the (re)constitution of self-identity (Giddens, 1991). By contrast, lifestyle is suggested as having less applicability in traditional cultures where the options in constructing one's life are more limited and identity tends to be 'handed down' rather than 'adopted' (Giddens, 1991).

Thus, Giddens' (1991) definition of lifestyle differs from Veal's (1993) in that it is focused on lifestyle as a vehicle for forming a more coherent sense of self. Accordingly, Giddens (1991: 81) defined lifestyle as 'a more or less integrated set of practices which an individual embraces, not only because such practices fulfil utilitarian needs, but because they give material form to a particular narrative of self-identity'. Semi-routinised practices, such as habits of dressing, what to eat, choice of work (or even not to work) and 'favoured milieux', can become 'decisions not only about how to act but who to be' (Giddens, 1991: 81).

In contrast to the privileged position of being able to choose among an array of lifestyles, some groups in various societies, of course, have less choice than others, with their way of life often imposed, rather than being a lifestyle choice (Veal, 1993). That being said, many who do have the resources to engage in other ways of life are not always aware of the range of life options available to them (Giddens, 1991). On that point, Giddens (1991) suggested that the more post-traditional the setting within which an individual operates, the more lifestyle concerns questions of self. Giddens (1991: 80) noted that 'by definition, tradition or established habit orders life with relatively set channels'.

Hence, lifestyle is often of particular significance to individuals engaged in 'alternative' ways of life that operate marginally to the 'norms', or traditional channels, of mainstream society (Metcalf, 1995). In view of that, significant research has been conducted into the lifestyles of groups such as surfers, hippies, ocean cruisers, ravers and rural communards (Macbeth, 2000; Malbon, 1998; Metcalf, 1995; Veal, 1993), and much of this research has been concerned with issues of self, values and perceived resistance. In Macbeth's (2000: 28) study of the ocean-cruising lifestyle, he observed that the adoption of a cruising lifestyle often reflects uneasiness about mainstream society, in effect a social critique, as many ocean cruisers have perceived that 'modern society is restrictive and saps personal choice and self-determination'.

In addition to ocean cruisers, which he describes as representative of a certain type of tourist, Macbeth (2000) also suggested that communal living, such as in Kibbutzim, and long-term travel may offer alternative views of how society might be constructed. Notably, each of the abovementioned groups seem to express a relative uneasiness with what Macbeth (2000: 25) described as 'the iterated structures of urban life and occupational imperatives of a career'. Likewise, Giddens (1991) urged that choice of work and work environment is fundamental to lifestyle orientations. So, it seems that much of the discussion on alternative lifestyles articulates a move away from traditional occupation-dominated lifestyles.

As lifestyle is largely constituted by the choices a person makes each day (Giddens, 1991), it has been strongly linked to degrees of freedom or choice (Veal, 1993). With long-term travel having been traditionally viewed as being counter to the norms of 'mainstream' society and characterised by a high degree of perceived freedom (O'Reilly, 2006), it can be justifiably considered as an 'alternative' lifestyle choice. Moreover, in most cases, the time commitment of lifestyle travel entails a move away from a career-dominated way of life. Lifestyle travel is a post-traditional context with its own ideologies and patterns of individual and group behaviour that are integrated not only at a functional level, but also foment a space where individuals may possibly seek coherence in order to try to 'make sense' of their lives.

## Conclusion

Through an examination of a historically constructed traveller/tourist divide, it has been shown that a traveller identity is largely borne out of tourist angst, wherein a traveller often self-defines/identifies in opposition to a tourist, her/his closest form. This self-definition through anti-tourism helps to cluster travellers into an identifiable group in its own right. Yet, there are also elements of travel itself, as a sub-type of tourism, which arguably allow for its differentiation. Travel itself tends to imply an extended journeying aspect, both outwardly and inwardly, that theoretically differs from mass tourism, which is often conceptualised as temporally constrained and perhaps, more playful. However, positioning travel as a form of secular pilgrimage as separate from mass tourism as a form of play is fraught with difficulties. For each individual's experience is just that, individualised, with elements of work and play that likely blur and change over time and place.

The burgeoning literature on backpackers, a label descended from the pejoratively positioned drifter and tramp, reflects academia's ongoing attempts to tease out the backpacker as a sub-type of the tourist. Backpackers have been characterised by the academe as journeying to multiple destinations for both longer than mass tourists and with a less organised itinerary and money per diem, all the while mostly utilising a backpacker infrastructure that has developed to capitalise on, and has reportedly homogenised, many backpacker needs. Yet, since its first use in academia, the backpacker label has been attached to the notion of youth (Loker-Murphy & Pearce, 1995), forming a stigma for some older self-styled travellers to resist, who can now potentially add backpacker angst as a subdivision of their broader tourist angst. While it is clear that labels such as traveller, backpacker and tourist are contested notions, it seems that through the haze there is a temporal factor that repeatedly emerges if one wishes to distinguish amongst these identities and within them.

Within the backpacker label, which has been traditionally aligned with Graburn's (1983) form of tourism as a temporary rite-of-passage, there exist a small proportion of travellers, as they like to call themselves (O'Reilly, 2005; Richards & Wilson, 2003; Welk, 2004), for which travel can no longer be considered a break or transition in their life span. For these lifestyle travellers, repeated and extended temporal commitment to a travel lifestyle, involving its own ideologies, praxis and identities, has become a way of life in itself that they may pursue indefinitely.

*Chapter 6*

# Backpacker Hostels: Place and Performance

MICHAEL O'REGAN

## Introduction

Backpacking, a form of tourism and sub-lifestyle, has often been placed in a very different category to mainstream tourism, a form resting on complex and interdependent infrastructural 'scapes'; producing (and being produced by) its own system of interrelated and increasingly interconnected institutions, transports, guidebooks, routes and symbolic spaces of consumption. This chapter takes a fresh look at the phenomena of 'backpacker hostels'; the network of backpacker-oriented accommodations, historically, discursively, symbolically and materially part of backpackers' lived 'socio-spatial practices'; part of a global system that enables, influences and shapes (and vice-a-versa) backpacker flows. The 'backpacker hostel' has risen symbolically and materially to become a validated and sanctioned portal for entry into this lifestyle, central to its reproduction and development, linking multiple spaces and times together; an important infrastructure and a key building block through which people relate to and associate with backpacking. More importantly, the 'backpacker hostel' is a place specifically for consumption and performance – routed in the discourse of spatial mobility, experience seeking, performance and identity. Celebrated and represented in film, media and literature as the antithesis to the 'International hotel', as the primary time/space experiential setting for a backpacking trip, it has become a key symbol of backpacker travel itself, where individuals perform, narrate stories, sample (or build) an identity, exchange knowledge and interact. As a key mobility system, a significant system of provision and a key consumption junction, this chapter traces the historic, symbolic and material meaning attached to them, emphasizing their role within contemporary backpacking.

## Conceptual Framework

We can conceptualize the 'landscapes of tourism' (Shaw & Williams, 2004) or more accurately, the landscapes of mobility as 'vacationscapes' (Gunn, 1989), 'leisurescapes' (Urry, 1990) 'travellerscapes' (Binder, 2004), 'tourismscapes' (Van der Duim, 2007), made up of various flows, which are constituted through a skein of complex, interlocking networks that increase and enable tourism between, within and across different societies (Urry, 2000). Within this framework, Urry (2000: 193) argues that there are 'networks of machines, technologies, organizations, texts and actors that constitute various interconnected nodes along which flows can be relayed'. This includes networks of transport of people by air, sea and road, as well as the wires, cables, satellites, fibre and microwaves that carry phone, email messages, images and money. The '-scapes' suffix signifies transnational distributions of correlated elements, a concept utilized by Appadurai (1990, 1996) to describe global flows, illustrated by the transnational arrangements of people, technology, finance, media and political resources, labelling them, respectively, as ethnoscapes, technoscapes, financescapes, mediascapes and ideoscapes. These 'scapes', Appadurai (1990: 296) argues are also 'deeply perspectival constructs, inflected very much by the historical, linguistic and political situatedness of different sorts of actors and provide for the foundation, the spaces and opportunities of "imagined communities" (Anderson, 1983)'. Appadurai (1996: 33) extends Anderson's concept of 'imagined communities' and argues that these 'scapes' are the building blocks of what he calls 'imagined worlds', that is 'the multiple worlds that are constituted by the historically situated imaginations of persons and groups spread around the globe' that constitute new forms of individual and collective expression.

For Bell and Ward (2000: 88) '[t]ourism represents one form of circulation, or temporary population movement', a privileged flow along these 'scapes' that is predominately structured and channeled by 'scapes' along different nodes, reconfiguring the dimensions of time and space (Williams *et al.*, 2004). As part of the 'ethnoscapes' or 'the landscape of persons who constitute the shifting world in which we live' (Appadurai, 1991: 192), tourists along with other moving groups, such as immigrants, refugees, exiles and guest workers, move around the world as global flows. Tourism is simply one form in a continuum of flows, helping to 'situate tourism in relation to other forms of mobility that are differentiated in their temporality and spatially' (Williams *et al.*, 2004: 101). Tourism is deeply structured by scapes – the very existence of motorways, flight routes, airports, etc., facilitating and channeling movement and are in today's globalizing world 'fundamental to understanding the massing of tourism flows along particular routes' (Shaw & Williams, 2004: 3). Backpacking, as a form of tourism, and a privileged flow (Alneng, 2002) has also been visualized as an imagined world, constituted, constructed and made possible through globalizing 'scapes', a differentiated flow or imagined community that has constructed (and is increasingly offering) its own version of these global scapes; a sub-lifestyle with its own routes, flows and rituals (D'Andrea, 2007).

It is through this social imagination then, that backpacking is often characterized as a distinct form of tourism, constructed as a more or less integrated set of practices and a role that an individual might embrace, an assertion that has seen them and their practices differentiated from 'mainstream tourists' (Cave *et al.*, 2007; Cohen, 1973; Uriely *et al.*, 2002; Welk, 2004; Westerhausen & Macbeth, 2003). They are characterized as taking up a different role within, between and across a de-territoralized 'landscape of scapes' – a 'travellerscape' (Binder, 2004), which describes the 'alternative' social arena or field in which backpackers experience their journey across, between and within borders and boundaries. The scape is constructed out of global scapes; anchored to a set of badges of honor, or an ideology; ordered 'according to certain sets of economic, political, ecological or social practices and discourses' (Ek & Hultman, 2008); with each individual backpacker participating and experiencing travel through a larger formation of like-minded individuals because of what such communion offers and represents. Bradt (1995) identified a number of ideological 'badges of honor', which included: travelling on a low budget to meet different people; to be (or to feel) free; to be independent and open-minded; to organize one's journey individually and independently; and travelling as long as possible. Pearce (1990) mapped out a similar ideology that included: a preference for budget accommodation, an emphasis on meeting other travellers, an independently organized and flexible travel schedule, longer rather than very brief holidays and an emphasis on informal and participatory holiday activities. Welk (2004: 80) believes that these 'badges of honor' are the 'basic symbols with which backpackers construct traveller identities and a sense of community' and 'serve to distinguish the backpacker from the (stereo)typical conventional tourist'. So, this imagined community in many ways constitutes many of the attributes of a lifestyle community, in the way they constitute themselves, their dispositions, their habitus as separate from tourists, as they orientate and embody, however slightly, these badges of honor or performative conventions.

While diverse mobilities come together to enable backpacking to happen, Urry (2007: 272) argues it is necessary to analyze the various systems that distribute people through time-space; given that mobility systems are 'organised around the processes that circulate people, objects and information at various spatial ranges and speeds' with all mobilities entailing 'specific often highly embedded and immobile infrastructures' (Sheller & Urry, 2006: 210). Furthermore, Hannam *et al.* (2006: 3) argue that '[m]obilities cannot be described without attention to the necessary spatial infrastructure and institutional moorings that configure and enable mobilities' and that no increase in fluidity can happen without extensive systems of immobility. Moreover, non-human objects, machines, times, timetables, sites, desires, systems, bureaucracies, technologies and texts provide 'spaces of anticipation' that enable a journey to be made, for a traveller to plot an itinerary and which 'permit predictable and relatively risk-free repetition of the movement in question' (Urry, 2007: 13). In this context, backpackers perform a particular version of mobility identifiable with the label 'backpacker', even though many who travel this way disassociate from the term as they buy into specific representations of particular spaces, routes, rituals and practices that have historic, economic, social, cultural meaning and significance (Hetherington, 1998). Backpackers' mobility-related performances, activities and practices thus do not exist in a vacuum, but rely on external institutions, infrastructures and systems, which are constructed materially and discursively to appeal to them specifically. Pooley *et al.* (2005: 15) thus argue that 'mobility is more than the mechanism through which mundane tasks are carried out' and '[m]ovement can itself become a performance through which we make statements about ourselves and acquire status'.

Backpackers, whose everyday conventions and practices require them to mobilize themselves spatially, intellectually, ideologically and culturally in new environmental settings, require structure – 'a degree of permanence, of fixity of form and identity' (Hudson, 2005: 17). Backpacker hostels are an important if not integral institutional infrastructure that enables, structures and represents their mobility. While Starr (1999: 377) suggested that studying infrastructure was the study of boring things, it is an important (even if sometimes mundane, unnoticed and embedded) part of people's lives, representing some of the most pervasive and foundational scaffolds of everyday social life. Hostels as 'infrastructure' and 'structure' have become a central point of reference within backpacker practices, providing a community and institutionally sanctioned portal by which means a socio-spatial practice is enacted, knowledge exchanged, where one's identity can emerge and be validated. While backpacker movement is not fully determined by hostels, it plays a major part within the system of interrelated institutions (transport, communications, roads, airlines) developed to support their mobility, and it remains the primary accommodation infrastructural network and a pre-eminent 'mobility system' for facilitating this global phenomenon. However, like mobility itself, the spread of hostels has been uneven. While some countries operate strict regulatory environments hindering their spread, others restrict foreign involvement and investment (by restricting labor, capital and knowledge mobility). Yet, even in those countries that contain a large, informal, heterogeneous accommodation base (i.e. Thailand, Bolivia, Peru and China), hostels are increasingly established.

There have been competing claims to the historical association between modern backpacking and hostelling. While Loker-Murphy and Pearce (1985) argue that the historical antecedents to the modern backpacker hostel can be traced back to the Young Men's Christian Association (YMCA) formed in 1844, most scholars see modern hostelling as having originated in Europe in 1909, when a German teacher, Richard Shirrmann, came up with the idea to take his students on excursions bordering the Rhine using schools along the route as accommodation (Clarke, 2004b). When Cohen (1973) investigated the nascent 'drifter' flows, precursors to today's backpackers, he noted they flowed along parallel geographic lines to tourists, 'institutionalised on a level completely segregated from, but parallel to ordinary mass tourism' (Cohen, 1973: 90). Cohen (1973: 97) noted that drifters sought out these cheap and conveniently located hostels, often called 'freak hotels', which acted simultaneously as lodging, meeting places and eating places, where 'youngsters exchange information, buy and sell their belongings, or smoke pot'. These original drifters, Cohen argued, travelled 'outside the established tourist circuit – both geographically and socially', making use of local opportunities for lodging, eating and travelling. Loker-Murphy and Pearce (1985: 824) noted how these developed into an infrastructure catering specifically to the drifters, comprised of 'inexpensive transportation systems, with low-priced hotels and youth hostels surrounded by psychedelic shops, nightclubs, and coffee houses'. Cohen (1973: 101) noted that 'drifter-establishments' were of low-grade and low-rate services, but still thrived on drifter tourism, but 'the ordinary caterer can expect little benefit from it' and in addition, 'the intrusion of the drifters into the itineraries and facilities used by ordinary tourists could spell a loss for the tourist establishment, since it antagonises the other tourists, for whom drifters are often anathema'. This led some researchers (Riley, 1988; Aramberri, 1991) to argue that backpackers were not concerned with their amenity surroundings or value-added services; a characteristic that meant that accommodation could be primarily offered by locals (Burns, 1999; Scheyvens, 2002a, 2002b), attracting backpackers in a bottom-up strategy of tourism development (Welk, 2004), where local people (primarily in developing countries) opened up their houses to relatively affluent nomads, forming nodes in a global ethnoscape and, as a consequence, were drawn into the multiple and disparate processes of globalization (Edensor, 2004a).

While the original drifter declined along with counterculture in the mid-1970s, recession and stagnation in the west; the spaces, narrative, memories, sights, sites and values associated with them lived on, revived in the form of backpacking in the mid-1980s. The 'freak hotels' and what they represented to the budget independent travellers were replaced by the backpacker hostel, a change made possible through massive growth in 'alternative guidebooks' and the travel media. These mediascapes, which refer 'to the distribution of the electronic capabilities to produce and disseminate information' (Inda & Rosaldo, 2008: 53) have further developed the new waves of budget travel and rather than being a marginal part of alternative literature, hostels have been increasingly reconstructed and conveyed through 'backpacker' films like 'Hostel' (2006), TV soap operas like 'Crash Palace' (Richards & Wilson, 2004: 267–268) and reality TV shows like 'Paradise or Bust' as central infrastructural scaffolds within backpacking travel. Strüver (2004: 68) argues that these representations 'might be carried in everyday speech, popular culture and high art, transmitted by TV, music, internet etc.' and lead to backpacking, their practices and consumption patterns being increasingly represented within consumer culture.

While hostels along the original 'hippie trail', such as Pudding Shop in Istanbul or the amir Kabir in Tehran, were famous in themselves made possible through word of mouth, hostels have become increasingly 'absorbed in the network', in which 'no place exists by itself' since its position and meaning are 'defined by flows' (Castells, 1996: 412–413) or more accurately, 'the inexorable speeding up of flow' (Broyard, 1982). This has led to a network infrastructure that is predictable and standardized even if not under single ownership, ensuring the same 'service' or 'product' is expected and 'delivered in more or less the same way across the network' (Urry, 2005: 245). Hostels, given their orientation, strive to tap into these global texts, resources and discourses, to legitimize their place in backpacker discourses, becoming '(re)entextualized' (Salazar, 2006) or reproduced, with some becoming better positioned as nodes; recognized within globally mediated texts, for certain performances. Visser (2004: 297), for example, notes that, 'the South African backpacker hostel sector seems to be emulating backpacker hostels in regions such as Australia and New Zealand with "first world" building designs, rather than using local materials and designs', rejecting the notion that separate geographic regions develop backpacker accommodations with their own preferred characteristics.

## Performing Hostels

Hostels are the most visible, material and symbolic part of backpacking culture, part of the backpacking script, a 'referential framework for the planning of a trip, but also a *script* for how to *perform* and perhaps reconfigure their own identities within the desired setting' (Jansson, 2007: 11; original emphasis). They have become so prevalent that some scholars argue that modern backpacking was born out and is maintained by backpacker hostels (Pearce, 1990; Slaughter, 2004). Wilson *et al.* (2007: 199) assert that Australia 'gained a competitive advantage in the global backpacker market because of its rapid and extensive institutionalisation and commercialisation of backpacker travel'. Backpackers and hostels are thus locked into a 'fluid self-reinforcing system' (Urry, 2005: 239) significant to those who pass through them and produced as a 'network' that enables embodied performances to occur. A backpacker hostel doesn't necessarily have to consist of similar people (age, gender, nationality), but of people sharing the same set of particular values, conventions, patterns of movement, involving intermittent physical face-to-face co-presence at locations on symbolic routes an important part of a network-driven community (Lassen, 2006: 307), even though individuals will not know exactly who will be encountered in these sites of 'informal co-presence' (Boden & Molotch, 1994; Urry, 2003). The decision to stay within these places of co-presence and communal proximity isn't 'incidental', but a conscious and habitual way of encountering and experiencing places and people, the symbolic nature of communal living differentiating backpacking from mass tourists; where 'inhabitants recognized each other, knew what they could or should do, and what relationships they could develop with each other' (Aubert-Gamet & Cova, 1999: 40). A hostel without other backpackers would seem odd given an audience is required to establish yourself, your road status and identity, the presence of others providing the reassurance that you are in the right place and reinforcing a 'commonality of experience that exists among fellow travellers' (Obenour *et al.*, 2006). Westerhausen and MacBeth (2003: 73) note the formation of 'such vibrant meeting places en-route' are a key component in that imagination, of chosen routes and ultimately destination choice, an imagined world that has become real, visible and negotiable (Römhild, 2002).

Rather than being mundane places of pause and transit, hostels enable rich, multilayered and dense interactions (Urry, 2003); shared space becoming an important conduit in the exchange process, whether it is the exchange of ideas, friendships, information and material goods, supporting a range of travel experiences from belonging, companionship, reflection and learning. It is also an environmental setting in which an individual can establish his or her own place, how to organize his or her time and his or her next move. While meeting other travellers may be secondary to meeting 'locals', it is still of significant importance (Obenour *et al.*, 2006; Cohen, 1973; Binder, 2004; Loker-Murphy & Pearce, 1985; Riley, 1988; Murphy, 2001, 2005; Richards, 2007) with many intense interactions forming far more quickly than they would in normal life, but also dissipating quickly (Elsrud, 1998; Riley, 1988; Murphy, 2001; Sørensen, 2003). Westerhausen and Macbeth (2003: 73) argue that 'like magnets in a stream of charged particles', sub-cultural meeting places have emerged. These 'gathering places' (Vogt, 1976: 36) permit backpackers to socialize with each other after traversing 'alien territory', serving to reinforce a communal ethos and contributing to the production of individual and group identities. Hostels have also been cast as a continuously updated travel advisory with relevant trip information, where information is exchanged, people put in contact with local events and available jobs, a place to make friends and organize temporary 'ad-hoc travelling groups to share costs, risks and experiences' (Binder, 2004: 98–100). These shared encounters are the glue of 'social networks and have a socializing effect in terms of mutual understanding, empathy, respect and thus tolerance towards others' (Willis *et al.*, 2007).

## Hostels in Development

While local accommodation development were 'blueprint beginnings' (Franklin, 2003), Cohen (1982b) noted in the early 1980s that on the beaches of Southern Thailand, where local ownership of accommodation was predominant, restricted access to capital made them vulnerable to being taken over by outside interests. Westerhausen and Macbeth (2003: 72) similarly note how '[t]he existence of flourishing backpacker centres frequently invites a "hostile takeover" of local tourism structures by outside operators and competing tourism sectors'. More recently, Brenner and Fricke (2007: 225–226) looked at backpacker development in Zipolite, Mexico, and found that 'developer-tourists' who expressly enter the market to build backpacker infrastructure have access to investment capital and business acumen, and so have a head start compared to the local population, gaining control over and dominating the backpacker market segment. While backpacking was traditionally seen as the first stage of tourism area life cycle (Brenner & Fricke, 2007), given the (perceived) market potential of 'thick flows' coupled with a global relaxation of foreign ownership rules, free movement of labor, objects, finance, technology, information, people, knowledge and capital; (tourist) developers, transnational companies and investment firms have the necessary capital, in-house (or hired) expertise and knowledge that is often beyond the reach of local entrepreneurs. Traditionally, while locals who were geographically tied to a locality entered to serve this market, outside firms and entrepreneurs, not surprisingly, followed this privileged flow, 'cherry picking' high-demand, low-risk and low-cost areas to try and maximize profits (Graham, 2004: 17). Clarke (2004b) found that individual capitalist involvement in Sydney's hostels acted 'in their own immediate self-interest', with the Sydney 2000 Olympic Games bringing multinational accommodation providers into the backpacker accommodation sector, a development that has seen both Accor and Starwood Hotel groups enter the market. According to Peel and Steen (2007: 1065), despite 'growing backpacker numbers, the continuing strength of local economic conditions, a booming property market and pressure to upgrade safety standards, the future of small independent hostels remains uncertain'. While economic interests penetrated and changed the original non-routinized and non-institutionalized character of drifter tourism in the early 1970s (Cohen, 1973: 95) leading to 'fixed travelling patterns, established routines and a system of tourist facilities and services catering specifically to the youthful mass-tourist' practice-specific economic infrastructure have now become embedded within this sub-lifestyle, as backpacker trajectories have become mapped.

The flow of backpackers has attracted an increasing amount of transnational investment both for profit making (Peel & Steen, 2006) and as part of local and national government strategies (Prideaux & Coghlan, 2006a). For destinations that are increasingly in competition in seeking to attract mobile capital or people (Hall, 2005a), the intimate relationship between hostel infrastructure and mobility in drawing international tourists to a region has attracted private investors and state funding. For example, a 2008 Suffolk University study found

Hostelling International-Boston's 32,800 annual guests pumped about $12.5 million into the local economy (Rivers *et al.*, 2008), while Clarke (2005: 315–316) notes how Sydney's Central YHA 570 bed hostel, which has its own travel agency, bar, convenience store, cafe, swimming pool etc., was politically welcome because of backpacker consumption in the environs of the hostel and because they are 'young, healthy, attractive'. Scheyvens (2002b: 157) meanwhile notes how 'the development of backpacker enclaves has transformed some run-down, crime-ridden parts of cities', given that backpacker are characterized as being young, fit, healthy, single, affluent and primarily but not exclusively white. In 2007, a public-private partnership with Hostelling International was announced as a centerpiece to revitalize downtown Winnipeg (Canada). HI regional director, Dylan Rutherford, said the area 'wasn't attracting the best crowd' – a case of public funding in serving a predominately non-local market. But backpackers are the 'right kind of transnationals' (Clarke, 2005); part of a vibrant cultural scene mix along with other previously marginal groups, a mix that attracts the 'creative class' (Florida, 2005), which in turn attracts capital. The ability of backpackers and the infrastructure that facilitates them to gentrify areas has also been noted in red light districts from Yogyakarta (Indonesia) to Sydney Kings (Scheyyens, 2002b; Visser, 2004; Howard, 2007), while pushing out marginal locals (homeless, drug users/dealers, the mentally ill). Increasingly, local, regional and national governments, through deliberate spatial strategies, are now seeking backpackers through the establishment of hostels, because of the capital they possess and also as a source of labor. Given that many backpackers would not visit a region without having the option of a hostel, the development of hostels are attractive both for filling seasonal labor vacancies as well as adding vibrancy to central business districts, which are normally devoid of night-time residents.

Australia, which has been short of labor for unskilled jobs, has traditionally not sought to fill such shortages with individuals from their near neighbors such as the Pacific Islands, but with individuals from countries thousands of miles away, facilitated by deregulation in transportation and communication systems. This type of working and holiday making is enabled by the federal government's 'Working Holiday Scheme' (Clarke, 2004a), which favors a very specific types of mobility – young, vibrant, cosmopolitan worker-backpackers (Williams & Hall, 2002), whom employers value given their enthusiasm and their mobility (Allon *et al.*, 2008), their search for an 'Australianism' (Morris, 2006) and encounters with the 'real Australia', making them a vital source of labor for seasonal harvest work and capital mobility. These 'harvest networks' often require up to 20,000 workers, the crop-to-crop mobility requiring hostel accommodation to attract and house them as long as the harvest producers require them. Hostel infrastructure along these routes play a vital function in this process by serving as the focal point for growers to source workers and post job information. These hostels act as agents, placing signs outside their establishments and advertising in back-packer magazines seeking harvest workers. According to Rural Skills Australia, farmers ring backpacker hostels requesting a specific number of workers, which the hostel will try to provide, with some hostels also supplying their own transport to farms, while others serve as pick-up points for buses that may be provided by farmers, harvest project contractors or local councils. The demand for these 'idealized mobile subjects' (Richardson & Jensen, 2008: 219) is increasingly important for places, highlighting the role of hostels in putting a place on the map, with local issues becoming 'more tightly and mutually intertwined with national and global concerns' (Edensor, 2004a: 11).

## Hostels in Enclaves

This chapter is not suggesting that the hostel network is at the cusp of being colonized by corporate giants, investment firms and profit entrepreneurs who, as agents of globalization, are offering a rich lifestyle for a 'relatively' cheap price, akin to the glocalized Starbucks network. It does, however, suggest that hostels, as an important consumption junction and a spatial setting for many practices that may dominate a backpackers daily activities, are taking on many of the characteristics of what Edensor (2000: 328) calls 'enclavic space', offering travellers a bit-part in a theatrical play. To encapsulate consumers, they create synergies with everyday activities (touring, nightlife, laundry, internet access); with particular consumption practices acting as 'binding agents' (Thrift, 2000) that give travellers enough autonomy so that they are seen and feel they are socially constructing, not socially constructed (Thrift, 1996), which reconfirms (genuine or illusory) individual autonomy (Bauman, 2001). Paraphrasing Edensor's (2004b) 'motorscape', hostels contain the com-fort of a spatially coherent identity, connecting places together in an increasingly enclavic-scape across differing local contexts with a familiar architectural style, containing familiar comforts such as wi-fi internet, English-speaking staff, security, live music, sociality, privacy – features that have become so common place and familiar within hostels, that they are now only noticeable when missing.

Creating a space for performance, also means restricting the mobility of others seen as economically marginal, immobile or disruptive, which usually manifests itself by refusing entry to locals even as paying guests (Hutnyk, 1996; Visser, 2003), meaning there is relatively little contact between backpackers and 'locals' (Obenour *et al.*, 2006; Huxley, 2007). Different bodies are separated out, transforming the travelling body in a series of processing categories, leading to spatial and racial segregation that excludes certain groups (stag parties, rugby groups, school groups) and ethnicities, which has recently seen Aboriginals in Australia, Israelis in India and even English tourists in Wales excluded from certain hostels. This exclusion extends to any individual or group who 'could' disrupt the enactment or performance of a backpacker's lifestyle or undermine the commercial activities of the hostel, while 'the other' act as staff 'performing' local culture. The ability of an actor to participate in the network is determined by whether they are seen as contributing to the goals of the network, denying entry to those who are unable or unwilling to perform supporting roles in the network. This also extends to individuals or groups that 'refuse to comply with the roles expected', those individuals who 'truly disrupt the stage and the normative enactions performed on it' (Edensor, 2000: 331), leading to their exclusion and alienation from the surrounding environment as hostels welcome similar people on a reciprocal basis throughout the world. Hostel owners will argue that this is their business, catering to international tourists, and that they have a duty of care to guard against intercultural misunderstandings by serving as a 'buffer against culture confusion' (Hottola, 2005: 5) on such issues as personal space, privacy, gender and sexuality. Hostels have become progressively more closed to the immobile (increasingly, hostels have maximum stay lengths), the non-paying or those considered economically insignificant. Surveillance, CCTV cameras, flow management systems, security guards and swipe cards help hostels offer 'assisted' and emplaced predictability and reliability – frictionless and seamless mobility through smooth corridors on the hostel network – a type of encapsulation, where the space is stage-managed, 'a strategy for maintaining spatial and imaginary boundaries' (Jansson, 2007: 9), an encapsulated spell that allows for an 'imagined' community to take hold, but 'in order not to break the spell, people are obliged to act in an appropriate manner – to play the right game' (Jansson, 2007). These hostel performances mean the boundaries of Self are not brought into question (Sibley, 2001), as hostels mediate between the backpacker and otherness, between edge and risk or any intrusion that might produce anxiety, while the individual backpackers them-selves, if threatened by others, can also retreat to private rooms.

Hostels are becoming increasingly adept at providing and organizing scripts, interconnecting with specialized transport companies and other backpacker operators through 'backpacker conferences', web-portals like hostelworld.com and specific hostel management forums (www.hostel-management.com) to manage a stage on which interaction will be carried out, modifying and manipulating spatial practices, tapping into global flows; helping create a scape that ensures travellers never feel lost, immobile or isolated no matter where they happen to be. Mobilizing shared activities and mutual aid systems that were traditionally tied to hostels, such as shared meals, kitchens, notice boards, wash rooms and communal chores become problematic in this co-production, as hostels, through accident and design, give new spatial meaning for more affluent mobilities. Many collective, shared practices and rituals that were once enabled by hostels are now withdrawn, no longer supported or inscribed within structures and are replaced by more individualized activities aimed at instant gratification, with less attention paid to activities that help self-organization or benefit a collective culture. Thus, the elevation of private space and commercial activities mean individuals can pursue their own interests, the '[e]xpectations of backpacker/hostel accommo-dation appear to be changing from the communal, cheap, "just a bed" option that it once was believed to be... something more in line with the accommodation experience of the mainstream tourist' (Cave *et al.*, 2007: 245). While the décor and layout of a hostel lobby once meant creating a very different experience from that encountered at a five star hotel, increasingly individuals are coming to expect very tangible and physical evidence of a servicescape (Bitner, 1992), whether it be internet access that works, to a level of service and performance from the 'management' and staff.

## Hostels in Play

Even though performances are increasingly prescriptive, individual travellers are active rather than just passive performers, producers as much as consumers. Jamal and Hill (2002: 100) note how tourists (as well as entrepreneurs and locals) can 'interpret for themselves and resist the ideological or hegemonic meanings being imparted by the industry or destination/attraction managers' and have the agency not to be co-opted 'to exercise performative freedom and resistance to being normal-ised into the dominant discourses' by escaping the institutions and structures that try and channel them (Sheller, 2001). Thus, backpackers will express themselves by fulfilling or occupying a consumptive role different from that of other tourists, positioning themselves in opposition to those 'conventional' tourists (Wilson & Richards, 2004: 123). Bruner (1991: 247 cited in Shaffer, 2004: 142) argues that this is often difficult given that 'once the tourist infrastructure is in place, the traveller can hardly avoid the well-trodden path of the tourists'. While Judd (2003: 34) argues that even when a tourist leaves his or her enclave to indulge in the unpredictable adventures of the flâneur, there is a 'limited range of options and choices available to them'. Yet, travellers will contest what is 'appropriate' activity in a given setting, engaging and engage in 'tactical revolts' and 'actively resist conformist performances' (Hannam, 2006: 244), not only sometimes arranging accommodation outside the network, but living outside the network geographically, materially and socially (Elsrud, 1998), committing themselves to the sub-lifestyle values by approaching the 'local' in a different manner (Loker, 1993; Riley, 1988; Scheyvens, 2002a, 2000b), evolving means and representation that they believe represents their social world more accurately than that which the industry can offer or impose.

Consumption might thus also be enacted through less privileged spaces, such as in heterogeneous spaces located to serve passing trade and the local population, accommodation that co-exists 'with local small businesses, shops, street vendors, public and private institutions, and domestic housing' (Edensor, 1998: 53), which may 'provide stages where transitional identities may be performed alongside the everyday enactions of residents, passers-by and workers' (Edensor, 2001: 64). These spaces are not connected to the global tourism industry to the same extent as enclavic space and do not (or may not have the abiity) to package and perform particular kinds of authenticity for backpacker consumption. These alternative spaces suggest or give the traveller (even if it is illusionary) the notion of serendipity, where a local place provides a convincing backdrop to identity creation. The traveller can perform when practicing language, for example, removing themselves linguistically and culturally while utilizing bargaining skills to find ego enhancement from getting 'best value' (Riley, 1988). Heterogeneous accommodation exists as rich and varied 'soundcapes' and 'smellscapes' (Edensor, 1998: 62) amongst the local shops, schools, hairdressers, markets, flats, residents, suburbs, street vendors and restaurants. In this context, Crawshaw and Urry (1997) similarly note the idea of a flaneur 'attracted to the city's dark corners, to chance encounters to confront the unexpected, to engage in a kind of counter-tourism that involves a poetic confrontation with the "dark corners" occupied by the dispossessed and marginal of a town or city, and to experience supposedly "real" "authentic" life uncluttered by the dominant visual/tourist images of that place'.

Backpackers are also deploying technologies by using social networking hospitality sites like bewelcome.org and couchsurfing.com, creating new forms of 'meetingness' (Urry, 2003); appropriated and lived as part of everyday travel experience, creating 'empathetic sociality' (Maffesoli, 1996: 11), overtaking the power exerted by institutions while creating new tactical media and responses in the face of massification. However, not all backpackers have the access, skills, competences, motivation and knowledge to move beyond the hostel network into heterogeneous space or other networks. Edensor (2000: 331) argues that most will acknowledge, accept and even welcome the controlled nature of enclavic spaces and adapt performances accordingly, 'prepared to trade self-expression for the benefits of consistency, reliability, and comfort' (Edensor, 2000: 331). For many backpackers, hostels are a reliable and habitual means to avoid the messy mobilities that one finds in heterogeneous spaces.

## Conclusion

The global processes enabled by technological developments in transport and communications combined with a wider 'appreciation' of independent travel has created and extended powerful (and affluent) thick flows that criss-cross the globe. This has embedded an increasingly interconnected, interdependent and mapped 'travellerscape', which includes a material and symbolic hostel network that serves as a 'mobility nexus' (Normark, 2006) – a place of identity making and identity habit (Jenkins, 1996), a distinct social space and a visible part of a sub-lifestyle. The hostel network is a symbolic space that connects places across time/space as well a material space that connects an 'imagined community' together through a shared movement of bodies, ideas, assumptions, stories and knowledge. Hostels are thus a key infrastructure scaffold, deeply embedded within backpacker travel arrangements; an infrastructure that supports backpacker mobility in a routinized manner. It is this global spread and habitualization that has led observers to suggest that these meeting places both represent a crucial component for destination choice for this form of tourism (Westerhausen & Macbeth, 2003), but also highlights its increasing institutionalization. Increasingly, hostels link together and with the media, transport, the state – offering a very consumption-driven community and forms of sociality, even if they only fleetingly represent familiarity and recognition, a shield of emotional protection, and a substitute for insecurity and unpredictability within the informal unorganized sector (Bell, 2005).

Backpacker hostels continue to reproduce and sustain themselves given they are an infrastructure considered as a foundational scaffold of everyday travel life – the primary form of accommodation associated with independent backpacker travel, symbolically, materially and discursively, providing an increasing level of predictability, safety and security. However, this is primarily achieved through exclusion, segregation and marginalization rather than by enabling any sort of 'local' contact or even low prices as they reterritorialize 'a sense of place' through theming and 'staged authenticity' (MacCannell, 1999: 92). Hostels have developed immensely complex 'participatory' programs to provide backpackers with consumption opportunities that are socially authentic, providing a stage for performances that tends to exclude those that do not want to or cannot participate, a strategy that increasingly mediates between the individual traveller and the 'outside world' whether it is for food, nightlife, laundry or internet access – a process that is welcomed by many. This process has led to a certain amount of de-individualization or reduced autonomy, a continual cumulative process whereby backpacking is captured and made safe for individuals, but only for those affluent enough to participate and willing to give up some individual agency for a 'controlled edge' (Hannigan, 2007: 73); a safe adventure and experimentation with identity that offers escape from everyday routines. Hostels then are safe ground for the individualizing and reconfirmation of self through encounters with familiar others but freedom and interaction is increasingly expected and purchased rather than earned as individuals interact 'lightly' without too much riding on the outcome (Hudson, 2005), escaping if boundaries of self are brought into question. While hostels increasingly become interconnected and interdependent through transport, informational and communications innovations, the scape by taking on a more enclavic nature might clash with individuals' imaginary embodiment in the world. Tactical responses, such as the use of heterogeneous accommodations and the use of the 'Technologies of the Self' such as hospitality exchange sites mean individual backpackers given the motivation, competence, access, skills and tools can reposition themselves along with others to coordinate their tactical movement through space on the 'promise of affection, conservation, a sense of new beginnings' (Turkle, 2008: 125). It may be that a new disjuncture is been created as individuals in coordination with others seek greater control of their self-image, as they set out to 'win space' that may be experienced as personalized, authentic, capital intensive and identity-enhancing. The question remains as to whether these creative and adaptive tactical responses and the creation of distinctly new social spaces can transform the global-local processes that have made hostels a key infrastructural system, acting as a preparatory step for geographically dispersed individuals to create new forms of alternative travel styles from below.

## Chapter 7
# Euro-railing: A Mobile-ethnography of Backpacker Train Travel

JAMES JOHNSON

## Introduction

The European cities of London and Amsterdam are considered as the original drifter centres of origin (Cohen, 1973). Yet, with few exceptions, the geographic spread of backpacking studies has tended to overlook Europe as a significant backpacker destination. Europe, however, remains popular with backpackers (Wilson *et al.*, 2008). Moreover, the opening of former post-Soviet states has brought a host of new destinations on to the backpacker travel itinerary, thus a return to the study of Europe as a backpacking location is needed. Backpacking around Europe is also closely associated with rail travel, not least through a series of inter and euro rail passes, and a tradition of making extended tours around Europe has been created and endures. Research needs to understand these transportational aspects of backpacking in Europe, not least because, with rail passes valid for up to 30 days, and with so many countries to visit, a significant amount of time is spent on the railways. This chapter is organised into five sections. The chapter begins with a review of the literature on backpacker enclaves and journeys, before going on to discuss work on train travel and transport mobilities more generally. The results draw from a 'mobile ethnography' carried out on board trains in central and eastern Europe during the summer months of 2007 and 2008 and attempts to form an under-standing of the role of the body in backpacker rail journeys; including the embodied response to the speed, conditions and movements in and around the rail carriage. Goffman's (1973) notion of body idioms is used to suggest that there are significant embodied interactions between backpackers as they travel. Conceptually, the chapter also draws from contemporary research into a new mobilities paradigm (Hannam *et al.*, 2006). The chapter explores the tactics backpackers use to create and maintain privacy in the presence of other travellers and within the compound of the public/private space of the rail carriage.

## Backpacker Enclaves and Journeys

> It appears that the great majority of the young contemporary backpackers spend significant periods of time, perhaps even most of it, in various backpacker enclaves, or on the road from one enclave to another... (Cohen, 2003: 98)

Backpacker enclaves have been described as spaces that facilitate meaningful interactions, allow for the communication of shared values and the expression of backpacker travel identities (Sorensen, 2003; Murphy, 2001). Enclaves are also a space of cultural contact for backpackers and local people (Teo & Leong, 2006; Wilson & Richards, 2008). Wilson and Richards (2008) argue that backpacker enclaves are 'dynamic' spaces where unique combinations of familiarity and difference, global and local cultures, work and play are created, 'suspending' the backpacker between a series of opposing forces. In these enclaves, entrepreneurial trade develops that builds from and feeds the desire for local and global tastes (see Ateljevic & Doorne, 2005). Enclaves are also 'metaspaces', capable of reducing culture shock, allowing perceived control and a respite from life 'on the road' and the unfamiliarity of destinations (Hottola, 2004, 2005). Hannam and Ateljevic (2007b) stress that although backpackers are sometimes relatively immobile in enclaves, such backpacker moorings are not separate to, but are enmeshed with a range of complex mobilities. Backpackers are often quasi-residents and move between roles of backpacking, work, study and play (Allon *et al.*, 2008). But, enclaves are not only points of interaction with local cultures and people, they are also sites of dynamic relations, contestation and power struggles, places where local and backpacking cultures can and often do collide (Richards & Wilson, 2008; Allon *et al.*, 2008). A system of surveillance exists where behaviours, activities and movements are scrutinised by peers, and are monitored or surveyed by locals (Desforges, 2000; Maoz, 2006). Enclaves are sites of backpacker accommodation, popular bars and cafes, and centres of communication that not only offer points of contact with other backpackers, locals and service providers, but also with the wider world and back home (Molz, 2006; Teo & Leong, 2006). Molz (2006) also speaks of a system of surveillance as travellers are emailing, checking, booking, posting reports and photos, effectively documenting their experience for others to see. In short, the backpacker enclave is a dynamic space that creates and allows for a series of connections between a range of people and facilities; enclaves are spaces of communication and activity; they are sites where backpacker interactions are played out and where backpacker experiences can happen.

Backpackers are often considered as highly mobile (Hannam & Ateljevic, 2007b) and, indeed, travelling is largely central to the backpacker experience. However, whereas there has been a great deal of recent research on backpacker enclaves, research into the actual *practices* of travelling to, from and within a destination has received less attention, an omission that is surprising when we consider that research on backpackers has traditionally been completed while on the move (Richards & Wilson, 2004; Hampton, this volume). The emphasis on enclaves seems to reflect broader tourism research agendas that have been fascinated with destinations (Coles & Hall, 2006). Indeed as Larsen (2001: 81) points out 'the significance of mobility or travelling to the tourist experience has been almost completely ignored in tourism studies'. Aspects of travel have often been reduced to transit zones within a 'tourism system' (Lieper, 1990). In the process, travellers are reduced to statistics and journeys reduced to departure and arrival times, echoing some transportation studies that have traditionally seen travel time as derived demand (Mokhatarin, 2005). Nevertheless, a strong, underlying rhetoric of travel underpins much of the backpacking literature and being a traveller or travelling forms a byword used by many backpackers to describe the type of tourism they are engaged in. Backpacker journeys have been underpinned by an almost heroic status (Riley, 1988) as the backpacker is a 'trail blazer' who ventures far off the beaten track (Cohen, 1973). Journeys are expressed as semi-nomadic, adventuresome and character building. Issues of corporal mobility and travel have been central but understated in many studies. Cohen (1973) originally conceptualised the ideal drifter type as one that *travelled* to remote localities. Vogt's (1976) wandering youth are described as having a 'travel style'. Mobility and a tramp-like wandering is the defining feature of Adler's (1985) youth on the road. However, as backpacker studies have themselves drifted away from anthropological towards managerial perspectives, a focus on understanding *travelling* has largely fallen by the wayside as studies have sought to understand backpacker decision-making processes and market segments (Ateljevic & Doorne, 2004). Yet, the rhetoric of travelling is still upheld. Pearce (1990), for example, describes contemporary backpackers as having flexible *travel* itineraries.

However, a mobilities perspective emphasises that we need to better understand practices that take place on the move. 'The mobilities paradigm posits that activities occur while on the move, that being on the move includes a series of occasioned activities' (Sheller & Urry, 2006: 213). Practices of travelling are enmeshed in a range of activities, experiences and interactions that have normally been associated with enclaves, creating a form of dwelling in motion (Urry, 2000). Studies of backpacking need to consider backpacker transportation as an integral part of the consumer experience and attempt to understand the activities backpackers engage in 'on the move' (Hannam & Ateljevic, 2007b). To quote Ateljevic and Hannam:

> Backpacker research within the new mobilities paradigm would thus examine the embodied nature and experience of the different modes of travel undertake, seeing these modes in part as forms of material and sociable dwelling in motion, places of and for various activities. (Ateljevic & Hannam, 2007: 255)

By separating out travel and destinations, the destination has come to represent the site of experience and activity – a discourse that places travel as an opposite, a necessary but boring evil, a means of getting from a to b in less than spectacular fashion. However, as Franklin and Crang (2001) stress, it is important to halt the separation of tourism from everyday practices, practices that embody cultural codes, ways of seeing, feeling and doing, and the more mundane activities tourists engage in. Tourism is a way of being in the world, part of everyday life and not an extension or an opposition to everyday life (Obrador-Pons, 2003). Seeing tourism as part of everyday life asks us to re-draw the boundaries of where tourism happens precisely because practice, actions, understand-ings and identities used and (re)figured when engaging in tourism stem from everyday understandings, or what Crouch (1999) describes as lay knowledge of the world around us. As Edensor (2006) argues – we take knowledge and habits with us on holiday, performances are often evoked by surroundings and situations, responses are habitual rather than reflective.

Recent studies on backpacker transportation have thus begun to discuss the more dynamic features of backpacker travel. Vance's (2004) conceptual framework on backpacker transportation choice discusses the characteristics and benefits of the transportation method chosen. Wilson *et al.* (2008), meanwhile, discuss the embodied travel experiences of New Zealanders on overseas experiences (OE) in Europe. However, to date, no study on backpacking has looked at the embodied practices of back-packers whilst on the rails in Europe. This study seeks to rectify this.

## Travelling by Rail in Europe

Backpacking around Europe is closely associated with rail travel, often through a series of inter and euro rail passes. A culture of 'inter-railing' exists in that backpackers seek out and actively choose to experience being on different European trains, an experience that typically includes extended train journeys across Europe. Often considered a democratic mode of travel, European railways have facilitated the ease of movement for large groups of people for the purpose of work and leisure since the 19th century (Urry, 1990). Rail travel signified capitalism's triumph over nature, forming an efficient method of transport capable of moving goods and people at increasing speeds along uniform, levelled-out tracks (Schivelbusch, 1977). For Virilio (1984: 40), rail travel is typical of modernity's quest to use new technologies to overcome barriers associated with distance and time, and he writes of a 'violence of speed' as technologies tear us away from the place travelled through. Classic accounts of the rail journey tend to support this view: 'rail road, I do not consider it travelling at all; it is merely being sent to a place, and very little different from becoming a parcel' (Ruskin, 1865 cited in Boorstin, 1965: 87). Boorstin (1965) argues that the inauthenticity and triviality of modern-day tourism experiences stems from the removal of the hurdles the traveller once faced. He argues that technologies like the railway have provided tourists with comforts where once there was trepidation and hardship.

It has been argued that, as technologies advanced, travellers became more and more detached from the natural and sensual aspects associated with travel (Schivelbusch, 1977; Virillo, 1984). Past modes of transport, horse-drawn carriages were held in close communication with nature; journeys were sensed, experienced and felt corporally, the hard wooden wheels, the swaying of the carriage, the crashing of hoofs, the feel of the lay of the land, bumps and all (Schivelbusch, 1977). Corporal hardships associated with travel were replaced with new and shocking sensations of speed as the body was projected through space. New ways of seeing the world developed based on this speed of movement and the technologies of the day thus included panoramas and the cinema screen (Lofgren, 2002). Larsen (2001) argues that the mobility of rushing landscapes and the immobility of passengers mirrored these new ways of seeing; like the movie goer, the rail traveller becomes a spectator, one who captures a series of moving 'travel glances'. Here, the rail passenger's body becomes reduced to a 'cinematic' metaphor, the training of the eye, the appreciation of fast-moving images, sitting, observing and finding comfort, almost docile, sedentary and penned into the almost perfect pantopia of the rail carriage, bound, held still, capable only of looking out onto passing landscapes (De Certeau, 1988). Movements of the body seem restricted by the physical structure of the train, controlled and regulated in the presence of other passengers, guards, the established norms and unwritten rules of rail travel. Interaction strategies aimed at reducing uncomfortable face-to-face relations are organised around the avoidance of eye contact. To be seen to uphold the rigours of polite company, objects and artefacts serve as markers of space and as shields to ensure interaction etiquette is upheld (Goffman, 1963). For these authors, the rail carriage becomes one of the non-places of Auge's (1995) supermodernity, a place devoid of identities and a place of fleeting relations, detached from the outside world.

Dann (1994) brings about an alternative reading by associating rail travel with the discomfort of long trips and the assault on the senses of being on a train. Lofgren (2002) talks of a tingling thrill of acceleration associated with experiencing new and faster modes of transportation. In short, the idea that as transport technologies advance, they become less of a sensual experience cannot be held entirely true. Viewing journeys through a cinematic metaphor overwrites the sensual aspects of travelling by rail, the embodied actions and emotions the passenger is likely to express and go through in the process of travelling means that the tourist body when on a trip is not merely textual; the body is sensually involved in surroundings (Obrador-Pons, 2003; Edensor, 2001; Crouch, 1999). In the process of making sense of doing tourism, everyday lay knowledges are drawn upon by tourists through a set of practices that are mediated through a sensual body (Crouch, 1999).

Drawing upon these non-representational ideas, however, does not undermine nor separate backpacker journeys from the more representational elements of backpacking. However, the train, the cabin and vestibule are sites of movement, relations and interactions. As the body is not cut from personal knowledge, neither is the backpacker's body cut off from surroundings, the presence of others, physical geographies and common codes of practice. Veijola and Valtonen (2008) demonstrate the role of the corporal body and the embodied and interactional aspects of being on a flight, including the gendered nature of embodied movements and interactions and the ill fit of technologies designed for a standard and idealised body. Thus, the body is physically and sensually as well as imaginatively situated in space and the traveller brings his or her cultural baggage and knowledge along for the ride (Edensor, 2007).

Moreover, research within the mobilities paradigm, particularly that of Urry and Lyons (2008), have recently linked rail travel with a series of embodied practices, such as gazing, reading for leisure, studying or working on the move. Here, mobile technologies provide valuable connections to co-workers, clients and friends. Symes (2007), meanwhile, shows how journeys to and from school allow children to re-create the internal space of the train carriage. Here, space is reconfigured into environments that support both play and homework. Complex interactions are played out, which are negotiated and regulated with and in the presence of other school children and other passengers.

## A Mobile Methodology

In terms of the research for this chapter, all data gathered was a result of extensive participant observation, interviews and logged conversations and, as such, the study follows the principles of ethnography. Research was carried out on board trains in central and eastern Europe during the summer months of 2007 and 2008. As I travelled, I had no set itinerary, preferring to follow the routes and travel patterns set by the backpackers I met, essentially allowing the field to determine the next journey taken. Field notes and observations were purposely very detailed and often completed in situ, and the results that emerged were thus somewhat fluid and dynamic. By moving with backpackers through journeys, the approach to data collection follows what Sheller and Urry (2006) term a 'mobile ethnography' as it involved 'participation in patterns of movements while conducting ethnographic research' (Sheller & Urry, 2006: 217). Mobile ethnography places an emphasis on 'travelling with people and things in a sustained relocation of the researcher within that movement' (Watts & Urry, 2008: 17). Mobile ethnographies are thus effective at detailing passengers, activities and actions as they are moved by and move in modes of transportation. Watts (2007) for instance used observations of train passengers over a number of journeys to critique accepted notions of travel time as wasted and provided a detailed analysis of passenger's spatial practices and activities. Similarly, Tucker (2007), by joining youth coach tours in New Zealand, was able to grasp the performative elements of such trips. This 'mobile ethnography' meant that 'respondents' changed on an almost daily basis in relation to settings and journey taken, the research emphasis being to record the actions, behaviours and norms of a range of actors and individuals

(Sorenson, 2003). As Sorenson (2003: 308) argues: 'The un-terrorization of the backpacker community means that instead of prolonged interactions with a few, fieldwork has to be structured around impromptu interaction with many'. Moreover, the changing of subjects in relation to journeys is an accepted and anticipated trait of researching those who travel by train (Symes, 2007; Urry, 2008).

## A Mobile Ethnography: On the Trains in Central Europe

> The train is busy there are four people in my carriage, Susan sits in the corner she stretches out her legs, as she does so the Michael opposite her leans back, she retracts. The younger backpacker next to me shifts his body from side to side, pulls the neck of his t-shirt and gives a gasp for air. A second girl enters, as she does so Susan stands and stretches she places her bag on the over head rack, freeing up space for another to sit. (Field Diary, 2007: Budapest–Split: 16/8/07)

Although seemingly mundane, banal and inconsequential, these aspects of journeying give rise to a series of questions about the movement of the body in backpackers' rail journeys. Corporal movements in and around the train carriage are like utterances, in that they are embedded and inscribed with meaning, forming a basis for interaction between backpackers as they travel. The excerpt shows that as a person moves from side to side, he or she shifts their body weight and there is a close tactile and haptic relationship between the feel of the heat inside the cabin and the sensing of the material surface of the upholstery. The backpackers' actions begin to externalise an inner desire to find comfort, providing what Goffman (1977: 31) describes as a body gloss: 'the individual pointedly uses the overall body to make otherwise unavailable facts about the situation gleanable'. As he or she does so, gestures that reflect private internal readings of the situation begin to transcend the body and enter into communication with the public domain, actions that others present can choose to ignore or respond to. The juxtaposition of the backpackers' bodies as they move in and out of shared space that neither lays claim to, reveal a close and embodied communicative interaction – a body idiom with those who share their immediate surroundings. As each move backward and forward, temporary territories of self are created, transcended and re-negotiated (Goffman, 1977). Thus, the body is rarely sedentary in rail travel, bodies are in constant motion, they are caught up in a series of interplays of interaction as space is shared and negotiated between actors. Small movements, or what Adey (2007) refers to as micro-mobilities, form a communicative process that go beyond the discursive limits of polite conversation, movements that are sometimes reflective, sometimes non-reflective and often automatically impart an inner reaction to the conditions of external space. Here, there is a fluid movement between what the body is exposed to – the feel and conditions of travel. There is a strong self-governing body politic at play as movements impact on and are regulated in response to internalised knowledge and the co-presence of others (Foucault, 1988). Issues of public and private space begin to take centre stage. For Goffman (1977), norms of acceptable behaviours are social sanctions that limit behaviours in public, private desires become subdued, adjusted as a means of keeping-up appearances in public. Slippages in gestures and utterances begin to call into question our ability to separate out what is public and private. Sheller and Urry (2003) bring about a more fluid interpretation by taking into account the moments that issues of publicness and privateness occur. For example, smokers often refrain from smoking in or close by open cabins. In other situations, smokers light up to the disgust and/or relief of other passengers. Some backpackers observed in this study, purposely lit up a cigarette whenever the train came into a station stop, at first as a means of deterring others from entering and later out of habit.

Urry (2008) draws upon Goffman (1977) to assert that interaction during rail travel is often based on impersonal communication and weak ties. Certainly, for some backpackers this holds true, one backpacker, for example, commented that she had 'spoken to lots of people but had not really met anybody – not like in the hostels'; often email addresses are exchanged with little intention of being used, suggesting such weak ties. In other instances, backpackers used the space of the rail carriage to facilitate communication and create ties, boarding a train in Serbia after a music concert, groups of backpackers were observed conjugating in the train vestibule, sharing drinks and stories of events, bodies of excitement ensued and a party atmosphere was created. Other backpackers also spoke of how chance meetings with other travellers helped to establish a bond and led to them travelling on together. In other instances, the presence of strange and suspicious bodies fostered fear and paranoia. The closed door of the cabin created a real sense of security; the shared experience established a bond between those travelling within and shaped a connection that at the time felt like a close tie.

As the journeys move on there is an unpacking and repacking of backpacks, private material items spill out and fill space of the train. The private use of MP3, mobile phones, hand-held games machines and the reading of books help to ensure moments of self-imposed privacy. Often, material items are used to lay claim to or act as markers of space, serving to privatise public areas, forming what Goffman (1977) terms the setting up of stalls. Possessions were laid out, towels and t-shirts hung out to air, items that are associated with a close communication with the body, extended the backpackers' personal space and encroached onto that of others. Backpacks were often left next to the owner as opposed to on the racks and these served as both a shield and a marker of space. Discarded and forgotten items, such as a notepad, a hostel reservation slip, a heavy oversized book, a banana skin and an odd sock left traces of past performances, reminding us that people arrive and leave in a (dis)orderly fashion.

Backpackers were often seen with their legs stretched across the cabin, their feet resting on the seat opposite. Such behaviour becomes assimilated into norms of behaviour while travelling. Couples lay against one another, comforting one another; these moments of intimacy and the open display of bodily contact demand privacy and create a sense of embarrassment and frustration for others. One couple commandeered a cabin to themselves by drawing the curtains and each sleeping across the three seats. Later, they reflected that a 'late one' the night before had left them hung-over, exhausted and in need of sleep. The movements of those outside were not visible to them and neither theirs to the wider public. By sleeping across three seats, they ensured that if anyone did enter they would be unlikely to remain. In other instances, the hot conditions led to visible signs of bodily discomfort that became difficult to control or hide, as backpackers' bodies sweated and gasped for air. Some attempted to regulate their bodies to uphold appearances in public. They would periodically disappear into the toilet to 'cool off' by splashing water over their faces or would stand close to the window on the vestibule to catch a cooling breeze. In more extreme situations, such tactical regulation becomes less possible as the body gives way to feelings of exhaustion, the body slumps, as the sun beats down relentlessly through the window and there is no escape. Bodies sweat profusely and movements become slowed down and reduced to the wiping of the brow, the pulling of the neck of a t-shirt and, in more familiar company, water is guzzled and poured over the body: there is no attempt to hide.

## Conclusion

This chapter has looked at the practices and experiences of back-packers as they travel on trains. The role of the journey has been considered at some length and it is suggested that travelling is a central part of the overall backpacker experience and certainly is more than a case of travelling to, from and in-between enclaves or destinations (Larsen, 2001; Cohen, 2003). Indeed, aspects of backpacker travel could be considered as an attraction in its own right. The ethnography suggests that backpacker travel provides opportunities for planning, meeting and interacting with other backpackers and also facilitates a range of experiences not too dissimilar to the activities associated within back-packer accommodation enclaves. Like backpacker enclave space, travel space is 'dynamic' and there are considerable overlaps in experiencing moments of travel and of dwelling. Using the train to sleep off a hangover or to meet new travel partners provide just two among many examples of how moments of moorings and mobilites seep into and inform one another.

Embodied practice has been central to this study and ideas stemming from Goffman's (1963) notion of body idioms as psychobiological embodied messages are considered as important elements in encounters. Here, movements and gestures facilitate embodied communications between backpackers; often in such communications the body can be purposely used to give off coded messages. In many other instances, body idioms are performed automatically with little before or after thought, but as a response to a specific situation. The body is also used to create and maintain space and this is often achieved through a series of connections between the body, knowledge and objects; for instance, by strategically placing backpacks or other objects on the adjoining seats to create more private space. The space of the rail carriage often takes meaning from the identities and performances of those within it and there are considerable connections between how this space is interpreted, used and created. There exist moments of self-imposed privacy, attempts to create privacy, attempts to create bonds and attempts to regulate self and surroundings. Often, too, movements and actions that start as reflective become assimilated into habitual performances as the journeys moves on and common practice sets in. Practice and performance then are not static and change over the course of the journey, bodily movements used to achieve comfort are constantly re-worked, and moments of being excited, scared, being comfortable and uncomfortable overlap and change from one moment to the next. The backpacker's body is not merely textual, overwritten by codes of practice, but also a vessel through which experience is felt and created. In other instances, practice and performance connect with what may be described as the rhythmic elements of train journeys. The backpacker's body connects with the stopping, starting and slowing and feels the bumps, the halts and speed of the train. As new people enter the space of the train cabin, bodily positions are altered and space is again reconfigured with new experiences occurring and new meanings becoming attached. Delays and unexpected halts to the journey often require passengers to get up, look, ask or at least wonder why.

*Chapter 8*

# Budget Backpackers Testing Comfort Zones in Mongolia

CLAUDIA BELL

## Introduction

This chapter seeks to extend the new mobilities paradigm that encompasses corporeal movement to include and explore the physical sensuous responses that tourists must adapt to as they travel. I refer to these spaces as comfort (or discomfort) zones; places in which the senses are satisfied or offended. The concept 'comfort zones' combines both spatial and sensual experiences. My investigation focuses on the flowing in and out of sensory comfort by budget tourists in Mongolia – expressed in blogs, and in my own ethnographic diary.

For many people, their own mobility is now an assumption: as technologies have accelerated, so have their own abilities to travel almost anywhere they wish (Bell & Lyall, 2002). For some, what was once 'a trip of a lifetime' has now become a way of life and for many tourists, that binary distinction between tourism and home life is recapitulated several times each year. As Hannam *et al.* (2006: 1) have noted, 'mobility has become an evocative keyword' for the 21st century. Through tourism, the Western consumer seeks a kaleidoscope of new experiences to add to the construction of self as an on-going project; to the assemblage of one's own autobiography through the purchase of travel events (Bell & Lyall, 2002). As Urry (2002b: 1) has argued, tourism 'is about consuming goods and services which are in some sense unnecessary. They are consumed because they supposedly generate pleasurable experiences which are different from those typically encountered in everyday life...'. These include diverse sensual experiences. In their introduction to the first issue of *Tourist Studies*, Franklin and Crang (2001: 13) called for tourism scholars to consider the novel experiences that enchant tourists and intrigue the range of human senses. Notably, they suggested that tourism academics consider the 'sensual, embodied and performative dimensions' of contemporary tourist practices (Franklin & Crang, 2001: 14).

I am addressing their call by pointing out that tourists' exposure to negative experiences might be used as a form of tourist cultural capital; experiences that claim even more authenticity than the sanitised version that is frequently encountered. This requires movement out of comfort zones, into areas that are more difficult, where the tourist has less control and where their normal daily rituals related to such activities as health and hygiene are modified, compromised or abandoned altogether. My case study is of budget backpackers in Mongolia.

Tourists have various agendas. Cohen (2004b) explains that most tourists want novelty and strangeness, but not total immersion in an alien environment:

> When the experience becomes too strange he (sic) may shrink back. For man (sic) is still basically moulded by his native culture and bound through habit to its patterns of behaviour. Hence, complete abandonment of these customs and complete immersion in a new and alien environment may be experiences as unpleasant and even threatening, especially if prolonged. Many of today's tourists are able to enjoy the experience of change and novelty from a strong base of familiarity, which enables them to feel secure enough to enjoy the strangeness of what they experience. They would like to experience the novelty of the macro-environment of a strange place from the security of a familiar micro-environment. And many will not venture abroad but on those well-trodden paths with familiar means of transportation, hotels and food. Often the tourist is not so much abandoning his accustomed environment for a new one, as he is being transported to foreign soil in an environmental bubble of his (sic) native culture. (Cohen, 2004b: 231)

Backpackers and budget tourists often seek more 'authentic' experiences. They are prepared to accept modest conditions in return for paying less and they know this may require a willingness to adapt to corporeal discomfort. This form of travel informs notions of their own identity: the self-narrative as the intrepid traveller gains credence. However, the trip is simultaneously their moment to perform heroically with the 'Other'. But even this group, ever eager for cultural currency, might sometimes find themselves in situations where the level of discomfort is ultimately challenging. As we shall see, this can certainly occur to budget travellers in Mongolia. In terms of the structure of this chapter, firstly I outline tourism development in Mongolia before going on to examine back-packer's tourism experiences there (including my own).

## Tourism Development in Mongolia

Mongolia has long been relatively isolated from the outside world. In 2007, World Bank researchers noted a recent spurt in economic growth, based primarily on mining and construction. They explained that low investment in other sectors:

> has been due to low returns – a result of costly and unreliable transportation services; lengthy and complex transit procedures, including customs and trade rules; distortionary taxes; coordination failures, at both domestic and international levels; and growing corruption. Poor financial intermediation is also a problem that has kept the cost of finance high, although lower than in previous years. Alleviating these binding constraints will ensure that Mongolia maintains the path towards sustained, broad-based growth. (Ian-chovichina & Gooptu, 2007: 1)

Tourism in Mongolia, meanwhile, has become a serious sector in the local economy only since the economic and political reforms of 1990. Today, many local entrepreneurs are trying to attract tourists. In common with other former republics and satellite countries of the former Soviet Union, the loss of economic protection from the parent super-power means that Mongolia now needs to invest in its own tourism facilities, infrastructure and skills base. Tourism is seen as a possible quick-fix economic saviour, though planning has been largely towards group tours, with an under-estimation of the contribution independent travel-lers can make (Baum & Thompson, 2007: 239).

From the late 1990s, visitor numbers have been climbing steadily, to reach over 300,000 in 2007. This has been aided by various internal strategies, including the 1999 Law on Travel and Tourism, which regulates tourism businesses and services, and the development of the Mongolian National Tourism Centre, which implements tourism policy (Buckley *et al.*, 2007). Various analyses on the progress of tourism have been offered (Hall, 2001). Saffery (2000) has addressed the role of local communities in Gobi tourism development, while Bedunah and Schmidt (2000) have studied the relationship between tourism and conservation in Mongolia. These studies centre mostly on infrastructural issues, including limited banking and credit card services, poor local transport, substandard roads, pot-holed pavements, overly-modest hotels and inadequate information services. Baum and Thompson's (2007) recent research, meanwhile, focused on the need for Mongolia's tourism industry to respond to the challenges in the local labour market, calling for significant investment in human resource development in Mongolia, which would in all likelihood require international funding. Similarly, Schofield and Thompson (2007) placed their research emphasis on local tourism management issues, producing evidence of an urgent need for more training for hospitality workers.

One of the few visitor studies available was conducted by Luvsands-vaajav (2005), a Mongolian researcher who surveyed tourists' impres-sions of Mongolia, and the implications for further development. Luvsandsvaajav's (2005) study asked departing visitors their impressions of the country, producing a quantitative analysis, which limited constructive critique of conditions and services. Yu and Goulden (2005), like Luvsandsvaajav, looked at tourists' satisfaction. They found that the landscape made by far the most dramatic impression. With the Gobi desert spreading across almost one third of Mongolia, over 40% of tourists visited the desert. Yu and Goulden (2005: 137) found tourists rated services and facilities significantly less favourably than other aspects of their tour (such as landscape, hospitality, cultural attractions). The findings of these surveys call into question the host's facilitation of tourists' needs. Yu and Goulden, and Luvsandsvaajav, saw their studies as offering a guide to further development in Mongolia. Yu and Goulden (2005: 134) suggested their survey results would be valuable to organisations and businesses in Mongolia evaluating their current operations and formulating their future management and marketing strategies.

## Why Go There?

The Danish adventurer Henning Haslund, who spent a large part of his life living with Mongolian nomads, started his 1946 book *Mongolian Journey* with a chapter entitled 'why did we set out?'

Why did we set out – I and those of like mind with me – for the world which lies beyond the limits reached by civilization's boldest cables and railway lines?...

There was so much that lured us – the desire to see what was hidden on the furthest of all known passes; the urge to test our qualities on tasks which were something else than just going on in a rut, as old people do; the longing to use young and untried muscles in a game which promised adventures and great experience... we started out in search of the adventurous world of our desires. (Haslund, 1946: 1)

Indeed, there is a great deal of travel literature about Mongolia. It abounds with a particular theme: travelling by camel or horseback for weeks or months or years with Mongolians, who are, after all, traditionally a nomadic people. A few contemporary examples include: *Mongolia: Unknown Land* by Jorgen Bisch (1963); *In the Footsteps of Genghis Khan* by John Defrancis (1993); *In Search of Genghis Khan* by Tim Severin (1991); *In the Empire of Genghis Khan* by Stanley Stewart (2002). In all of these texts, the Mongolian is firmly positioned as 'other': fascinating, exotic and always far less civilised than the visitor. Travel writing as a commercial genre offers engaging narratives that fulfil readers' quest for accounts of the 'other'. In short, the secure self is tested as he/she must manage unforeseen contingencies, mobilising into and out of comfort zones.

In *Wild East – Travels in the New Mongolia* (2002), the journalist Jill Lawless, who lived in Ulaanbaatar for some years, described her journey through the Mongolian countryside. It had not occurred to her that there may not be roads. Mongolia is vast, but has just 750 km of paved road. Her vehicle averaged a very bumpy, uncomfortable 20 km per hour, with the driver chain-smoking. Then it broke down. The driver grabbed various tools and made a fire on which to heat and bend bits of metal. The travellers sat on the cold ground, worried about spending the night in the open, in that vast, unbounded space, while it was getting dark. The author gathered dried dung to make a fire – for her, another new experience. Then the Mongolian driver drank quite a lot of vodka. She reflected that, here 'was a glimpse of the Mongolian character: self-sufficient, hardy, stoic, resourceful. No wonder the Mongols won an empire' (Lawless, 2002: 43). Next day, when the vehicle still would not budge, the driver led the way to seek help at a distant *ger*. Hours later, Lawless was welcomed by an old woman, who gave her dried curd to eat, a mug of salty milky tea, with tough greasy mutton, then fermented mare's milk vodka. She wrote, 'it shot down with a sharp vodka tang then settled with a disquieting dairy after taste'. Later, the visitors walked a mile to wash at a well. Indeed, all travel writers who visit the Mongolian countryside remark on the generous local hospitality. Strangers are welcomed into homes, offered food and may stay the night if they wish. However, Lawless contributes another perspective:

I'll let you into a secret: there's not a foreign resident in Mongolia who hasn't secretly dreaded that famous nomad hospitality. And I'll tell you why: it's the food. The essence of the Mongolian diet can be captures in one word: mutton… a grindingly unvaried diet; boiled mutton with rice, boiled mutton with noodles, boiled mutton in a dumpling. Mongolians have what the Lonely Planet guidebook referred to as "one of the world's most rudimentary cuisines". (Lawless, 2002: 169)

Lawless describes the fatty, salty weak tea and fly-dotted sour cream. She is polite, wishing never to affront her hosts, but recounts being served a chewy stew with a 'thick slab of fat the size of a paperback' resting on top. Furthermore, at restaurants in Ulaanbaatar, she noted the dim lighting 'designed to prevent you seeing the hairs, gristly bits and mystery chunks in your dinner' (Lawless, 2002: 170).

## Researching Travellers Going Out of 'Comfort Zones'

The primary methods of collecting material for this research were traveller's internet blogs, and auto-ethnography about my own experiences as a budget backpacker tourist to Mongolia. Arguably, the new travel writers in the contemporary world are the bloggers and there are numerous blogs about travelling in Mongolia. In these accounts, written in diary or travel journal form, the contemporary tourist is a self-styled amalgamation of explorer, photo journalist, anthropologist and diarist. These writers cheerfully recount their corporeal discomforts while in Mongolia. They also add more qualitative detail than those tourists whose views were quantified in Luvsandsvaajav's (2005) paper on perceptions of Mongolia as a tourist destination.

Unlike traditional private diaries, blogs are written to be read by both familiars and strangers. 'By personalizing content, blogs go beyond an informative role… bloggers can present their ideas as they wish' (Bruns & Jacobs, 2006: 4–5). However, any form of diary-keeping is a minority habit. Blogs are used as research documents cautiously; the behaviour and attitudes of diary and blog writers cannot be proven to be either typical or unusual (Alaszewski, 2006: 117). Nevertheless, the multitude of blogs available offers the qualitative investigator extremely convenient primary data, with no need for any temporal or spatial synchronisation between the researcher and the subjects. However, allegedly factual material may be unreliable; plus the public nature of this story-telling makes its content selective. But, for the researcher seeking individual unsolicited travel stories, blogs are something of a goldmine. Bloggers can say exactly what they wish about a place they have visited, even if that might be insensitive to their hosts. As we shall see, in Luvsandsvaajav's study, the respondents appear respectful, with very few commenting on poor drinking water, low levels of hygiene or difficulties with food. Yet, these are key topics of conversation for tourists. Certainly on internet blogs, the discomforts of travelling in Mongolia are a frequent motif.

In the following section, I also address a range of discomfort zones commonly encountered, and experienced by myself, in Mongolia, with reference to Reisinger and Mavonda's (2005) categorisation of travel anxiety perceptions. Auto-ethnography challenges any notion of academia requiring 'silent authorship, where the researcher's voice is not included in the presentation of findings' (Holt, 2003: 2). As Duncan (2004: 8) asks, 'for qualitative researchers, willing to confess that reality is based on perception, why should we not examine what constitutes our perceptions?' Auto-ethnography makes particular sense in researching tourism topics as it supports that adage that you had to be there.

Indeed, I was in Mongolia – for a few weeks in 2006. During this time, I kept a detailed hand-written diary, illustrated with drawings, local newspaper cuttings, purchased postcards and other ephemera. I noted observations, conversations and personal reactions. I did not turn the travellers (myself and my photographer partner) into heroes; indeed, I recorded discomforts, exasperations, illness and diarrhoea, in defiance of the traditional 'everything is lovely' travel journal. I considered the diary both a comfort zone and a personal database. There was also notable correspondence of my own experiences with some of the published travelogues, and with that of the bloggers.

## Discussion

In the Lawless (2002) account, her journey to the countryside involved constant delays, waiting about for a vehicle to be repaired, misunderstandings with the driver and anxiety at being forced to camp over night. In my experience, our old Jeep broke down, too: we were often asked to climb out and push it. We found this amusing, rather than disturbing. However, I was worried when I watched our driver consume a large quantity of fermented mare's milk. Lawless readily admitted to her own anxieties, of often feeling vulnerable and stressed. Reisinger and Mavonda (2005) write of how perceptions of potential risks influence choices to travel. Those risks might include potential life-threatening events such as terrorism; so, most prospective tourists will choose another destination. But they refer also to the 'risk' of not having control of situations, and at having one's general autonomy and well-being compromised. They also address the financial risks that tourists encounter, when they feel fiscally on edge, ripped off or not provided with fair value for money. The published Mongolia travellers' tales referred to earlier, illustrate the embracing of risk and possible danger. For those travellers, waiting around for a few days for something to happen is simply part of the journey. They were little interested in assurances about their own safety, and prepared to move into risky situations for the sheer excitement and exhilaration of it; many back-packers to Mongolia have similar attitudes.

We quickly became aware of everyday hazards and discomfort. Our rental apartment in Ulaanbaatar was in an eight-storey block housing perhaps 800 families. The sealed triple-glazed windows without fire escapes engendered a sense of entrapment. Thick dust, mites and *kapok* made us cough. Daily, we left our apartment through the treble-locked steel door, bracing ourselves against the reek of old boiled mutton, stale cigarette smoke and sewerage emitting from the adjoining apartments. The caretaker lived beneath the grimy stairs, surrounded by stinking garbage. This was collected daily, but the stench lingered. All night, the racket of steel-cutting on the nearby high-rise construction sites re-minded us that the summer building season is short; so building takes place round the clock with floodlights. Some tourists tolerate such daily discomforts, in exchange for a genuine local experience.

Finding a driver and guide turned into an extremely expensive mission. Like Lawless (2002), we had not considered the lack of roads, or the vehicles' needs for constant maintenance. As a result of thwarted expectations, and the waiting around for information, a lot of budget backpackers hang out in Ulaanbaatar, either waiting for visa extensions or for their flight out. They had not taken into account that the Gobi is a 2000-km round trip; or an expensive air flight away, which needs to be arranged well in advance, and with a group. And when they get there, some are disappointed that the desert is *not* golden sand dunes, like the Sahara.

We hired a driver, Jeep and guide to take us to Terelj National Park. Such romantic ideas about staying in a small *ger* camp, just the two of us in our cosy tent, enclosed from the elements: a honeymoon *ger*! We were surprised when the driver and guide shared the little tent with us. (We also learnt that foreign tourists stay in *gers*; Mongolian holiday-makers prefer cottages or huts.) A heavy torch was provided. We were advised that if we went outside to the toilet (bushes) at night, this would be useful for bashing wolves. This socio-cultural experience attacked our comfort-zone boundaries. We certainly had no fear of our companions; we had simply envisaged another more private scenario. At the same time, we were aware of the Mongolian tradition of sharing personal space; the assurance of greater safety through having these men as, effectively, bodyguards; and the shortage of finance to separately house various visitor groups and staff.

Local food quickly becomes a theme for any short-term tourists exchanging stories about Mongolia. In Western culture, we are taught to accept any hospitality, especially food, with grace. This can be challenging when the local diet transcends familiar food taboos; and when its smell, taste and appearance are counter to what is generally acceptable at home. While apparently polite to their hosts, the bloggers later confessed their actual reactions in the comfort zone of their blog sites:

> In restaurants, you'll be eating alongside the flies. Out of UB, there is little running water, so the routine goes a little like this... pick up dung, put on fire to heat oven, handle raw meat, cook, serve to horrified customer. http://www.dooyou.co.uk//food/mongolia-dishes/413493

> We were served our dinner which looked like pastries filled with something. I was assaulted with a layer of oil and a foul smell. The others ate the pastries "to be polite". I thought I was really polite to try it. After discovering it was organ meat, I ate a tiny bit more. Sylvie started to feel sick and then got worse when we discovered the *ger* next to us housed drying meat covered in maggots. http://www.mego.net/rtw/blog/category/mongolia

> Tea is made with horse milk which has fatty globules on the top and little gopping black bits. It tastes like the smell of a dirty horse, and it is terribly rude to refuse it. http://www.bintmagazine.com/bint_stories/906.php?story_id + 464

> You need a hardy stomach or time to adjust to the high fat content and subtle (limited) palate. Whilst I gorged on goulash and booz, every other traveller I met (without exception) was stricken with indigestion and stomach maladies. Ignore the smell of mutton fat and dig in! http://www.dooyou.co.uk//food/mongolia-dishes/413493

As the various commentators above noted, dealing with Mongolian food can sometimes be challenging. The smell, lack of hygiene in its preparation and service, and source of raw ingredients all contribute to the diner's reactions. My own observation was that:

> A Mongolian in a restaurant eating identifiable whole pieces of boiled offal off a sheath knife, out of a large chipped enamel basin, belching, spitting, as globules dripping down his chin, contravenes

all Western ideas of etiquette, appropriate food and propriety. The mutilated sheep carcass lying on the driveway in a puddle of dried blood outside the window, and the lack of hand-washing facilities for staff and guests at that same restaurant, plus the one over-flowing toilet, further disquiets the anxious tourist.

This kind of experience can take the tourist a long way out of his/her comfort zone. Deborah Lupton (1996) reminds us that the experience of embodiment is socially produced, and food and eating practices are always mediated through social relations. Lupton (1996) refers to Fischler's 'omnivore's paradox': the search for new foods to taste, coupled with the sense of danger or anxiety about the unknown: 'The incorporation of the wrong type of substance may lead to contamination, transformation from within, a dispossession of the self' (Lupton, 1996: 17). Western travellers have traditionally defined boundaries around foods deemed edible and inedible; which as Lupton notes, are closely related to other common binary oppositions, such as self/other, inside/ outside, good/bad, culture/nature. When food norms are transgressed, the response is often an emotional one, but – politely – at an unarticulated visceral level. Disgust, distaste, abhorrence, repugnance: tourists have the good sense to identify these feelings within themselves, and know that it is inappropriate to express them. Food arouses emotions, because eating usually involves sensual pleasure. As Lupton (1996: 29–31) explains, there is a very 'strong link between taste and smell, and the emotional dimensions of human experience'. There is also a strong relationship between memory and the emotional dimensions of food; we often recall events via the sensations of taste and smell. Indeed, backpackers in Mongolia frequently recall the food as the following extract from my ethnographic diary concurs:

And then there is the famous incident with the dead sheep. This crops up frequently in tales by travelers who visited the Mongolian countryside. During my own visit to Mongolia, it went like this: Out in the countryside, some-one ran to tell us they were about to slaughter a sheep. We, being squeamish city people, are quite happy for them to do this, but somewhere quietly, please, out of sight. Our host, aware that tourists enjoy spectacle, invited us to watch the sheep being butchered. He appeared disappointed when we declined. Later he ran back to urgently tell us "they remove entrails now! You want to see; maybe help?" Well, no, we didn't want to do that, either. We watched the eagles circling, and drank beer while small groups of locals gathered to watch the sheep butchery. We felt like spoilt-brat tourists, there for our own ends (including food), but not willing to involve ourselves with this essential everyday task. This expression of local culture identified us as distinct outsiders.

We were aware, everyday, of the juggling of emotions, and the transitions between unpleasantness and enchantment. Our own experiences, and others we heard about – rooms with no lights, broken-down overflowing toilets, smashed doors, inebriated taxi drivers, illness and diarrhoea, street-kids harassing in Sukhbaator Square, drunks at street markets or surprising offers from prostitutes – all became part of the tourists' memories and narratives. These are events that certainly 'intrigued the senses' as Franklin and Crang (2001: 13) suggested, though sometimes distastefully; and also contributed to the tourists' sense of their own heroism. Tourists went there, had a great time and stoically survived conditions they would not tolerate at home. They were aware of local poverty, and the limited resources to feed tourists; but issues around hygiene exacerbated discomfort.

## Conclusion

The experiences recounted here contribute to the 'emotional' and 'sensual' turn in tourist studies, demonstrating that tourism is another venue that mediates sensation. It marks 'an appreciation of the interconnected location of emotions in people and places' encouraged by Bondi *et al.* (2007). It also affirms the hegemonic role of travellers, as they respond to the way of life and local conditions of the 'other'. This chapter also extends the new mobilities paradigm that encompasses corporeal movement, to include and explore the physical sensuous responses that tourists must adapt to as they travel. Budget backpackers in Mongolia appear to be on a quest to 'immerse the body' in new contexts, unimaginable a generation earlier (Hannam, 2008). The pursuit to experience 'authentic local culture' may mean travelling far out of comfort zones. These comfort (or discomfort) zones are places in which the senses are satisfied or offended. The travellers described here indeed 'participate in their own skins' (Franklin & Crang, 2001). Backpackers attracted to locations associated with discomfort or risk take in their stride any crisis, angst or danger that makes for a good story later. This is all cultural capital for the backpacker.

Moreover, the valiant backpacker frequently places the Mongolian as an exotic 'Other': the 'Other' is positioned in a manner that reinforces the expected hegemonic or neo-colonialist travel narrative. The bloggers subtextually affirm Western identity in contrast to what they experience in Mongolia. The extreme climate, dusty accommodation in high-rise buildings, street markets awash with dirty water and rubbish, unfamiliar food based on boiled mutton and horse milk and limitations of the embryonic tourism service sector: these all require tourists to move beyond their usual comfort-zone expectations, and adapt to the actual experience of Mongolia.

*Chapter 9*

# Lesbian Backpacker Travel Experiences in New Zealand

LINDA MYERS

## Introduction

In the contemporary world, many women have a great opportunity to travel both during their leisure time and through their business activities. They may choose to visit places of interest to them and can occupy both public and private spaces en route. This chapter considers the literature on gay tourism and the scant literature that considers lesbian women's travel. The chapter then explores in detail the motivations, personal development and meanings of travel for lesbian women backpackers through their journeys and experiences in New Zealand. The research draws upon qualitative data from 25 in-depth interviews with international women travellers conducted over a three-month period during the New Zealand summer of 2006–2007. Two themes emerged, in particular, and are examined in detail: escaping the heterosexual world and social constraints, and lesbian identities and spaces.

## Lesbian and Gay Tourism Research

According to Hughes (2006: 3) in his book *Pink Tourism*, 'academic studies of gay and lesbian tourism are limited' and 'material relating specifically to lesbians is very limited in quantity and coverage'. Furthermore, Waitt and Markham (2006: 33) acknowledge that, 'very little academic literature exists about gay women and lesbians in terms of travel and tourism'. For example, a recent critical analysis of gay tourism, edited by Clift *et al.* (2002), considered a wide range of topics that investigated the relationships between sexuality, place, identity and meaning. However, much of this research concentrated on the homosexual male with a particular interest in their sexual behaviours. As Paur (2002: 4) noted, 'The focus on men's travel, to the exclusion of women's, is both an historically entrenched problem and a failure to incorporate gendered analysis into conceptualisations of tourism and travel'. Indeed, it is only recently that gender, let alone sexuality, has been seriously examined in relation to tourism (Kinnaird & Hall, 1994). Academic studies of tourism in general have focused on the male (and hetero-sexual) traveller, with 'such emphasis having produced hegemonic, disembodied and masculine knowledge' (Johnston, 2001: 180). Moreover, Jeffreys (2003) concluded that lesbians have a set of interests, culture and history of their own and must not be confused with those of gay men. Hence, Pritchard *et al.* (2000: 275) acknowledged in their study of sexuality and holiday choices that, 'in the same way as there seemed to be differences in the nature and expression of sexuality between lesbian and gay men there appeared to be distinct differences in how "they" (the participants) define lesbian and gay holidays, there was a definite sense that lesbian holidays are very different holidays'. They define a typical gay male holiday as involving 'Nude bathing; lots of partying and drinking; sunbathing; trying to look good and sex. That's a summer holiday', but by contrast, 'it proved to be much more difficult to define what, if anything, particularly constituted a lesbian holiday' and that they 'were more difficult to define precisely because they lacked strong associations with sex and the body that characterized many gay male holidays'. They found that '[r]eflecting the lack of a distinct lesbian identity in tourism, the women in the study felt poorly served by the current position of travel information guides and travel agencies. The gay travel industry was felt to be much more attuned to the needs of men' (Pritchard *et al.*, 2000: 275). Perhaps one of the only recent research papers to focus specifically on lesbian travel has been Kanta's (2002) paper about lesbian tourism to Eresos, on the Greek island of Lesbos and Sappho's birthplace, containing ethnographic accounts of the seasonal lesbian community that is re-created every summer, a community with its own territorial and symbolic boundaries, numbering over 1000 women.

Nevertheless, questions regarding the relationship between sexuality, space, and tourism are gradually being addressed and are no longer considered taboo in the wider academe. The main focus has been the importance of leisure and tourism in the expression of gay and lesbian *identities* (Hughes, 1997; Aitchinson, 1999; Clift & Forrest, 1999a, 1999b; Pritchard & Morgan, 1999). Hughes (1997: 3) maintains that 'the taking of holidays by homosexuals serves a role in constructing identity'. Waitt and Markwell (2006: 6) state that 'like the pilgrim, the gay tourist often seeks to experience "magic" that is enhanced by "group identity"' and they concur with Howe (2001) that 'same-sex tourism is like a pilgrimage, a quest for individual and collective identity'. The tourist experience provides the opportunity to access 'gay' space and temporarily abandon heterosexual norms, and gives the opportunity to 'come out' albeit for a short period of time. Pritchard *et al.* (2000: 267) claim that gay people have long travelled for recreation and often to escape intolerance and argued that 'the need for safety, to feel comfortable with like-minded people, and to escape heterosexism – often to specifically gay spaces – emerge as key influences on their (gay and lesbian) choice of holiday'. Hughes (1997: 6) concluded that '[g]iven that society has discouraged openness about being gay the holiday provides the perfect chance to come out, if only temporarily. The gay identity can be adopted and confirmed in "secret"'. Evidence in this research suggests that women do search out specific accommodation hideaways in order to fulfil their need to identify with other lesbian travellers and hosts.

Moreover, from a more managerial perspective, in her assessment of gay men and lesbian women's hotel experiences, Poria (2006: 330) concluded that the awareness of staff towards same gender guests and the tactful assignment of bed types were influential factors in guest satisfaction. Participant responses noted that 'the bed is a symbol of their identity and sexual orientation' and that 'bed arrangements might be the only way to recognize whether their sexual orientation is accepted'. According to Truco (2004: 2), the term gay-friendly refers to gays and lesbians being fully accepted by service providers and '[t]hat means no one bats an eye when two men or two women request one king-size bed. The concierge should be knowledgeable about gay activities, events, neighbourhoods and nightlife'. The hotel guest mix was also a factor influencing hotel experience because it affects gay and lesbians freedom to behave as couples in public places and their opportunity to 'perform being gay'. Roth (2001: 273) commented that '[a]t least during our vacation, we should be free to be ourselves in a welcoming environment'.

## Gay and Lesbian Tourism Infrastructure in New Zealand

The *Lonely Planet New Zealand Guide* (2006) devotes only three paragraphs to gay and lesbian travellers, acknowledging the low profile status of this niche market. It states that local New Zealanders are fairly relaxed and accepting of homosexuality and points out that there are prominent gay communities in Auckland and Wellington. It states that '[t]he biggest excuse for a party is the huge HERO festival held every

February in the Ponsonby road district of Auckland'. Other national venues include 'Out Takes', the name given to a gay and lesbian film festival in Auckland, Wellington and Christchurch in June, whilst Queenstown stages its annual winter Gay Ski Week. There are no specific lesbian events, but lesbians are included under the gay umbrella. For further travel information, it recommends the following internet sites: *Gay Tourism New Zealand, Gay Travel New Zealand, Rainbow Tourism* and *Gay Queenstown.*

Indeed, the internet is increasingly becoming the means of attracting lesbian and gay visitors to New Zealand. *Gay Tourism New Zealand* works very closely with New Zealand's gay and lesbian tourism industry on gay and lesbian tourism-related issues and supports and endorses the industry's 'Gay/Lesbian Friendly' and 'Rainbow Tourism Accreditation' (formally 'Approved Gay/Lesbian Friendly' criteria) for mainstream tourism operators who have expressed a commercial and marketing interest in being involved in the gay and lesbian tourism industry. They state that: 'Over the last 10 years, New Zealand's gay and lesbian tourism industry has grown from a network of gay and lesbian owned home stay's scattered throughout the country, to a cohesive network of accommodation ranging from home and farm stay's to luxury lodges supplemented with a network of boutique hotels to rental car and coach services' (Gay Tourism New Zealand, 2008: 1).

New Zealand was the first international gay and lesbian destination to adopt the 'Approved Gay/Lesbian Friendly' banner in 1998 and use the term 'gay/lesbian friendly' for straight tourism operators wishing to become involved in New Zealand's gay/lesbian tourism industry. This branding was necessary, as New Zealand law requires no discrimination based on sexuality or lifestyle. The branding was also necessary to maintain a quality 'gay/lesbian friendly' tourism product, so that New Zealand could be promoted as a genuine international 'gay/lesbian friendly' destination to the gay and lesbian traveller.

## Methodology

For this research, 25 women travellers were interviewed about their travel experiences in New Zealand. A semi-structured interview method was used and each interview lasted approximately one hour. The interviews took place during the 2006–2007 summer season in New Zealand between December and March and were conducted mainly at women-only hostels, lodges and campgrounds. Where possible, relaxed, quiet, informal surroundings were chosen to conduct the tape-recorded interviews. Respondents were predominantly European in origin and a high percentage were very well educated. In addition, respondents were subsequently asked to provide photographs that symbolised and represented their holiday experiences. After returning home, approximately one third of respondents replied with both photographs and sometimes summative comments about their travel experiences.

In terms of who's travelling, I found that there were solo lesbian women and lesbian couples travelling together, ranging from 19 to 75 years, two thirds of whom were over 30 years of age. Of course, there are no official statistics collected in New Zealand that require visitors to declare their sexuality, however, one lodge owner who had been in the lesbian tourism market for many years stated that:

> There are three distinct groups of women travelling, the young gap year mainly from Europe – now becoming the thing to do much like our OE (Overseas Experience). Second group are in their late 20s–early 30s these are the women who are wanting to self discover. They are often removing themselves from a difficult relationship, a break-up, a career change or some other personal trigger. Then there's the 3rd group those with disposable income 40–70 year olds who want to discover other cultures, scenery and enjoy travelling to gain knowledge and experience. (Rose, aged 62, from Australia)

The research evidence collected in this study does appear to generally back up this observation. The women in this study were highly educated, mostly to degree level, articulate and very willing to give their opinions. Over half of the women interviewed had taken time out from their professional occupations.

The following sections report the findings from the interviews. Two themes are explored in turn: escaping the social constraints of the heterosexual environment and the search for space – both personal space as well as the space to share and interact with other lesbians. In this context, Wearing and Wearing (1996) have offered an alternative conception of tourism not as an escape from the everyday, but as an escape to spaces that foster interaction and self-development: '[t]his space is akin to Foucault's (1980) concept of "heterotopias" or spaces outside of everyday life that allow individuals to resist dominant discourses' (Wearing & Wearing, 1996: 8). Both conceptions, escaping from everyday life and escaping into an alternative space, apply to these research findings based on lesbian women's experiences.

## Escaping the Heterosexual World and Social Constraints

The lesbian women interviewed expressed their journeys and experiences as *escapes* relative to the societal expectations and perceptions in their home environments. They wanted to escape from the roles they played at home and to have the freedom from socially expected behaviours to examine their own social and sexual identities.

For example, Andrea, aged 46 from Germany, argued that:

> at home I know everybody in the countryside so I, I, I can't be myself, only in my house. In this country at the other side of the world, they tolerate lesbians more than in Germany and it makes you feel free, much more free to kiss her and hold her hand in the outside.

Similarly, Sabine, aged 32 from Germany, noted that:

> it's better in New Zealand to be gay I think than in Germany, I get more people, erm, yeah, being rude to me because of my butch looks of being gay at home, it's better here there more accepting of who I am.

The acting out of multiple identities was a common attribute of these women, however it should be noted that only two interviewees had declared themselves 'out' to everyone in their home environments, but this was the exception rather than the rule, allowing them to develop their identities in a less fragmented manner. The complexities of lesbian lifestyle for the individual are illustrated by the following.

Anne, aged 55 from England, stated that:

> At home you wear many hats, that of a partner, a daughter, some yeah em as a mother (but not me), a sister, niece and so on, oh and an employee or neighbour. If you can imagine, sometimes my hat displays the lesbian rainbow flag well when I am with a partner and if I am with my sisters, or other lesbian friends but the rest of the time the ribbon is hidden and I have to conform to the expected role/s of a straight woman. When I go on holiday I wear my rainbow hat lots more and that allows me to relax and be me.

The anonymity of travel seemed to empower lesbian individuals and give them freedom to perform their personal desires. Helen, aged 30 from England, argued that:

> Nobody knows you here so you can do, em, weird or strange things, ha, ha, and no-one will think oh what's she doing? Or something, em,

yeah anything can happen, the worlds like open now and, your not confined in it, like in a box, you can be whoever you want to be actually and, em, you get more em, em, I don't know, you do the things you want to do instead of things people want you to do.

Lesbian women are escaping the structures and associated constraints of home life in two ways: of being a woman in a heterosexual world and of being homosexual women in a heterosexual, often homophobic world. This perhaps generates an even stronger desire to carefully select a holiday destination or environment that will allow at least some 'relative escape' from these constraints. The significance of a defined lesbian space should not be underestimated. With reference to the only commercial women's space in Manchester's Gay Village, Pritchard *et al.* (2007: 284) argue that the café *Vanilla* 'provides a place to drink and escape the pressures in the rest of the village but it is developing into a centre through which sports, holidays, excursions and nights out are arranged. These extended support functions seem to emanate from the isolation of lesbians in society as a whole'. In New Zealand, the lesbian-owned café at Waipu village, Northland, provides a similar service and is a place where lodge visitors are taken to so that they too can feel part of a wider lesbian community. The owner argued that the café was a place:

> where they can feel safe to be themselves and be in a friendly environment, both the village and in our community up here. Where they can be part of a bigger lesbian community. They're more able to show their affection just because of the construction of the environment erm, they don't have anyone watching them, you know, defining how they can be. (Cara, aged 42, from New Zealand)

Whilst a guest, Andrea, aged 42 from Germany, pointed out that:

> the café is a great meeting place, just seeing other New Zealand lesbians out shopping and calling into the café for their daily socializing was lovely to see, they were welcoming to me as an overseas visitor, I felt part of the community something you don't experience at home, very special. You looked forward to going again to meet more lesbians and like they just accepted you as a fellow lesbian.

By escaping heterosexual socio-cultural norms, a freedom is gained that can result in a sense of empowerment. A sense of inclusion as opposed to exclusion can be experienced. From Foucault's (1980) post-structuralist point of view, power should not be thought of as only repressive (as it is for lesbian women in patriarchal, heterosexual environments), as it can also be an empowering and motivating force (overcoming societal structures to live, albeit for a short time, as they choose). In this case, temporarily living and interacting in a lesbian travel space, hidden from home societal structures can result in a sense of freedom and empowerment.

Often in their home countries and environments, the women interviewed did not go out especially to seek public lesbian space, often because they found that, firstly, such space was difficult to locate especially outside renowned city spaces, such as those found in Manchester and Brighton in the UK and Hamburg in Germany and, secondly, because most 'gay space' is dominated by gay men, which is not necessarily appealing to lesbian women. The women interviewed, on the whole tended to socialise in small groups in their own private spaces at home, or go out as a group of lesbian friends into heterosexual spaces such as pubs, restaurants and night clubs. Therefore, the opportunity to perform the lesbian role in a relaxed and safe environment was very limited, making the importance of public lesbian space in the travel environment even more important and special.

Lisa, aged 43 from Australia, noted that:

> mmm, I mean at home we actually don't seek out either women's space or lesbian space except that our best mates are women and lesbians, its very difficult to find anyway, well I suppose we get around that because we shut the world out really and we visit one another's houses where we can talk about lesbian topics and be affectionate with one another in front of our friends.

The women interviewed negotiated and escaped the heterosexual world by creating their own small, secluded or 'secret spaces' in which to socialise in their home environments. Whilst away from the home environment, they became more anonymous and this afforded a greater sense of freedom with regard to their sexuality. In addition, they actively searched for places and spaces whilst in travel mode where they could meet and enjoy other social lesbian company.

## Lesbian Spaces

The lesbian women expressed their journeys, escapes and experiences as relative to their limited spatial freedoms, which resulted from the socially constructed heterosexual world. Women generally negotiate constraints (Wilson & Little, 2005) by actively searching destinations, places and spaces, where they gain some power, control and freedom to perform and express their identity. As Prichard *et al.* (2007: 276, 285) argue, '[g]ay and lesbian places are empowering places, providing men and women with a sense of community and territory' and 'have emotional and psychological importance as empowering places in a "straight" world'. For example, Karin, aged 26 from America, argued that:

> it's just, it's just relaxing to be around lesbians and you don't have to explain yourself it's, yeah, it just feels normal, I feel, yeah it gives me power, it gives me energy sort of and okay as long as I'm single of course I always get a good look at all the Kiwi girls.

The interviewees wanted to escape the heterosexual environment and share female space with other lesbians and women to re-affirm and perform their femininity and sexual identity. Myslik (1996: 157), in discussing gay spaces, comments that, 'these are spaces that enable not only open displays of intimacy and affection but also provide access to a variety of gay and lesbian friendly services and facilities, including shops and bars'.

Similarly, another respondent Vera, aged 40 from Finland, noted that:

> I'm always interested in how women live and especially how lesbians live and I'm interested in the country and travelling too, erm, so it's also a little bit of time out, have a rest, to reflect my life at home and it's important for me, I mean I, I need the, the exchange with other lesbians, with other, erm, queer identified people, I like the culture and want to feel part of it.

In this context, Hughes (1997: 4) has argued that, 'a sexual identity is partly an individual construct but it has to be validated by others, both homosexual and heterosexual'. Lesbian space in cafes and different accommodation types that were visited during this research in New Zealand created opportunities for individual lesbians to validate their identities with others.

Ceri, aged 58 from New Zealand, commented:

> I always stayed in women's places and places run by women if I possibly could, erm, so, and you do lots of talking so it was all really important and, erm, in the end being part of the women travel network now, you know, we're quite self-consciously sit and identify with women and when it's appropriate to as lesbians.

The Kiriwana annual lesbian women's camp, located in the south part of North Island, provided a further 'lesbian space' in which to perform being lesbian and validate social and sexual identity in a safe environment. Visitors shared the camp with its associated chores and participated in loosely organised and relaxing activities, such as basket weaving and stone carving. Both domestic tourists and international tourists, including some well-established couples and single lesbians congregated in the camp for one week, creating a uniquely lesbian albeit temporary community. This environment was totally segregated from the heterosexual world and its associated constraints and expectations and, as such, provided a blank canvas upon which to validate lesbian identity, celebrate lesbian identity and create a truly lesbian space and community.

The owner/organiser, aged 64 from New Zealand, enlightened the researcher to her philosophical views behind the lesbian camp as follows:

> Travel is about developing the inner mind, ones inner energy is affected by external happenings. By experiencing the camaraderie and friendship of other lesbians here in the camp hopefully we can together positively enforce our lesbian existence.

Sabine, a camp visitor aged 32 from Germany, reflected that:

> at first I felt a bit scared as I didn't know what to expect but I am amazed how quickly I relaxed. The conditions are basic, the loo buckets over there and so is the shower bag in that tree. Somehow we all seem to gel, I suppose we have being lesbians as the main thing in common and we just emm share the jobs, like gathering wood and herbs and cooking. I really like the welcome flags and banners at the gate they were good hey yeah Its like a relaxed sort of community developing, I like it, I feel I belong.

and she went on to say:

> I think the best part so far being here is last night, you know around the camp fire, I loved playing my guitar, and getting people singing but mostly I liked being able to hold my girl around the shoulder and kiss her when I wanted to, that felt good, you can only do that when you feel comfortable with your surrounding people, you know what I mean.

Spatial freedom and the associated relaxed atmosphere in the camp gave participants the opportunity to converse with one another about lesbian topics, strike up potential relationships, be affectionate with partners and develop feelings of pride in being lesbian. The researcher observed that the creative activities led to 'expressions of being lesbian', such as carving a woman's form in limestone and a gender symbol in pumice, as a further means of expressing and sharing identity. A quiet peaceful atmosphere whilst working allowed for long and more intimate conversations to develop. Favourite lesbian singers and other lesbian celebrities and icons were talked about, demonstrating the need and intention of sharing the common binding lesbian knowledge.

In this context, Sharon, aged 51 from New Zealand, stated that:

> Being here has made me feel how out of place I feel at home, I was married but it didn't feel right, I feel much more like a lesbian here and part of a world community somehow, do you understand.

Aggie, aged 51 from England, further commented:

> I think the strong and long established lesbian identity of the two women providers here at the camp creates an accepting atmosphere in itself, a good foundation for developing your own positive feelings about being lesbian and of course associating with all these other lesbians from different cultures with their own cultural stories and differences to tell.

And Jennifer, aged 55 from England, noted:

> I think you get the best out of travel by not moving on quickly by taking your time at one place. I am so enjoying being here for a week, although it is basic accommodation it is tolerable because you go past the superficial meeting of people, I have had some fantastic, lengthy conversations here with the hosts and the other lesbians staying leading I think to a much richer satisfying experience.

One of the most common conversations in the interviews revolved around the booking of accommodation in a heterosexual world where the pull of conformity and societal expectations had to be negotiated and confronted. There were numerous stories about booking (by whatever media) 'the double bed' and on arrival having staff presuming a mistake had been made during administration, once they are confronted by two women together. The following quote from Danielle, aged 72 from Scotland, demonstrated this well:

> we just arrived to stay in this hotel and the guy calls across the lobby as I, after I check-in coz he's apologising, you know, shouting across

the room that he's terribly sorry he didn't know that we were two ladies and, golly horror, you know, he's given us a double room, he'll change the room immediately so I shouted back, you know, no it's fine, you know, it's absolutely, you know it's fine that's what we want and he just looked at me and (once the penny had dropped) he said 'oh, well that's alright with me then', ha, ha [laughing with embarrassment].

Frequenting lesbian-advertised or lesbian-owned accommodations, however, overcame any such predicament and allowed for greater relaxation and less judgmental situations to develop. Hospitality staff, trained to have an open mind and be sensitive to such issues, can improve the experiences of the lesbian guest. As Poria (2006: 331) argued, an important finding in her study was that 'gays and lesbians wish to be treated and given the same experiences as they perceive are given to heterosexuals' and that 'gays and lesbians want to feel that they are accepted as they are, and in the hotel context, this means mainly being able to sleep together and be treated as a couple'. Similarly, as an accommodation provider, Ceri, aged 58 from New Zealand, argued that:

They're more able to show their affection just because of the construction of the environment. Yep. Well, and also because they're in that women's space where they can practice being themselves and don't have anybody watching them or expecting, you know, defining how they can be. When on holiday they can have a go and do something different.

However, my respondents also had a clear sense of needing personal space as well. They wanted to create what they referred to as '*me time*' in beautiful, relaxing, peaceful and safe surroundings, to explore their own self-identity. For example, Corrine, aged 26 from Switzerland, argued that:

I was like really stressed out, I, I had a lot of trouble at home and a broken heart and, yeah, what I expected was like find myself a little bit and get peace and quiet again and figure out what I'm gonna do, ha, ha.

The use of the words '*beautiful*', '*peace*' '*safe*' and '*freedom*' were commonly used by the women in relation to their needs and desires. New Zealand was perceived as a destination that provided for those needs in particular, as Yvonne, aged 37 from Holland, noted:

I'm sort of in a mid-life crisis and what I need now its just peace, quiet and beautiful, natural surroundings, its just I hope I get some time to think and, em, be a bit closer to me instead of always thinking of others.

Escaping the roles in the home environment of daughter, grand-daughter, aunt, mother, friend or colleague, allowed a greater focus upon 'self', self-exploration and hence personal development (see Myers & Hannam, 2008; Wilson & Little, 2005; Jordan & Gibson, 2005).

## Conclusion

This chapter has produced an insight into lesbian backpacker tourists visiting New Zealand and has been contextualised with reference to the literature on gay and lesbian tourism more generally. It has explored the identities of a group of lesbian travellers and attempts to understand the potential that their travel experiences have had in enhancing their self. In the 21st century, many women in the global community, particularly from the western world, now have the opportunity, freedom in society and financial independence to choose to travel for pleasure to other corners of the world. Additionally, in some societies, women have chosen to develop a lesbian identity and as such, search for other lesbian company and lesbian space during their leisure time as a means of expressing this identity. This chapter thus attempts to add to the body of gay tourism research by developing an understanding about lesbian women and their motivations and desires in a tourism context.

During the interview process, the discussions at times caused the women to reflect on themselves; this self-realisation in itself became a learning process for both the women and the researcher. New Zealand was chosen as a destination by the interviewees because it had specific provision for gay women in the form of accommodation and tours, it was a location that offered freedom, peace and safety and it was perceived as being gay friendly. Group or collective identity was a significant feature of the travel experience for the respondents. Interestingly, the lesbian women were generally not in search of a new partner or for new sexual experiences, as is common with gay men whilst on holiday or travelling, but were more interested in a quest for friendship and a space to be themselves.

The lesbian women were searching for 'lesbian space', free from heteronormative pressures; a place to interact with other lesbians, make new contacts, friendships, new networks and a space in which to explore their own identity. Those couples interviewed felt it important to have a public space where they could be what they termed 'normal', where they could be affectionate, without being subject to the 'normal judgmental gaze' with its associated expectations. Achieving this position in a lesbian-only accommodation then became an empowering experience. In this case, temporarily living and interacting in a lesbian travel space, hidden from home societal structures can result in a sense of freedom, empowerment and can validate identity.

## Chapter 10

# Backpackers as Volunteer Tourists: Evidence from Tanzania

KATH LAYTHORPE

## Introduction

The concept and evolution of backpacking since the 1960s has been well documented (e.g. Cohen, 1973). It is a contentious term as backpackers often consider themselves 'real' travellers compared to the conventional tourist (Welk, 2004). Clearly the backpacker scene, over the latter part of the 20th century and the start of the current one, has been dynamic; changing from the early idea of a hippie counterculture to one that has growing significance in the global market (Macbeth & Westerhausen, 2001). Welk (2004) suggests that this process of change has led to a fragmentation of the original concept, thus making it difficult to 'determine its boundaries from other forms of tourism'.

Backpacking is characterised as a travel style that emphasises freedom and mobility (Ateljevic & Doorne, 2004). Cohen developed a typology of tourist roles that made a distinction between the institutionalised and the non-institutionalised traveller (Cohen, 1972, 1973, 1974). Ateljevic and Doorne (2004: 62) suggest that institutionalised tourists are seen as cocooned within a 'Western tourist bubble' in which decisions are made for them and their needs are met within the tourist infrastructure, whereas the non-institutionalised traveller is considered to be searching for 'novelty, spontaneity, risk, independence, and a multitude of "off the beaten track" options'. However, by 1982, Cohen had already begun to articulate the change, in what was to be later termed the backpacker movement (Pearce, 1990), into a more conventional tourist model.

The contemporary backpacker is considered by Hottola (1999: 79) to be resistant to such typologies, as his study of women backpackers in India and Sri Lanka identifies a number of segments within the market, focusing on the heterogeneity of those who backpack. The contemporary backpacker is, therefore, now seen to embrace a myriad of identities

(Ateljevic *et al.*, 2004). Nimmo (2001) observed that for some backpackers there was a desire to step outside of their consumer role so that they may, more readily, become immersed in 'authentic' socio-cultural experiences. For these backpackers and those researched by Uriely and Reichel (2000), opportunities were afforded, through volunteering, to enjoy extended periods of social engagement. Little research has been carried out into the juxtaposition between volunteering and backpacking, however. This chapter seeks to begin to address this through research conducted into the lived experience of volunteer tourists who were also backpacking either prior to, after or more interestingly, *during* their volunteer placement. This, also, raises issues of the interplay between work and leisure, especially as the volunteers are termed 'volunteer tourists' who may, subsequently, backpack within the space of their work placement – perhaps a holiday within a holiday – depending upon how the concept of volunteer tourism is defined.

As with backpacking, research on volunteering remains in its early stages. Volunteer tourism is, however, starting to draw the attention of researchers and marketers alike (Wearing, 2003). But despite the rapid growth and the popularity of this type of tourism, Brown and Lehto (2005) suggest that 'systematic academic research in this field, particularly from the perspective of the volunteer vacationers, is still in its infancy'. Volunteer tourism is seen by many as a means of gaining a more authentic experience or as Brown and Lehto (2005) describe it 'cultural immersion'. Volunteers, it appears, are able to gain first-hand experience of another culture and to become more physically and emotionally immersed in the local community. They believe that volunteer tourists have the opportunity to move beyond the typical tourist platform, to see people 'as they really' are and therefore gain a more authentic experience. Indeed, as McIntosh and Zahra (2007) suggest 'it is unlikely that other cultural tourists will gain the same depth of interaction and experience as a volunteer tourist'.

Brown and Lehto (2005) state that preliminary research suggests that volunteer tourism can be sub-divided into two different forms depending on the volunteer's mindset. These two types are 'volunteer-minded' and 'vacation-minded'. The volunteer-minded are those who spend most or all of their time doing volunteer activities, whereas the vacation-minded spend only a small part of their holiday on voluntary work at the destination – 'brief encounters that have often proved to be the highlight of the individuals' vacations' (Brown & Lehto, 2005). As the authors reflect, this classification scheme takes on a simplistic view and certainly there are also clear examples of more complex interactions between work and leisure taking place. The erstwhile volunteer can, for a period of time, become the traveller/backpacker during the course of their project. The classification becomes even more complex as the volunteer tourist can move from a position where their main motivation is quite genuinely altruistic to one in which leisure activities begin to become more attractive and to take up an increased proportion of the time – a moving closer, therefore, to a more vacation-minded model.

Following on from this, Callanan and Thomas (2005) begin to draw attention to the heterogeneous nature of the volunteer tourist in their classification. They use the terms 'shallow', 'intermediate' and 'deep' to describe the different range of volunteers, and within this classification they begin to try to categorise the characteristics of those who volunteer. Their framework identifies that some participants are long-term volunteers who are on their projects for longer than six months, whereas for others the time spent in their placement may be less than two weeks.

## Methodology

Between May 2007 and November 2008, three visits were made to the Kilimanjaro region of northern Tanzania, comprising, in total, 18 weeks. On each occasion, the researcher stayed at a volunteer house in the region. Over 100 volunteers were observed and 40 semi-structured interviews took place, each lasting approximately one hour. However, of those interviewed, only about a quarter fell into the category of volunteer/backpackers, but the information gained from the other interviewees also contributed significantly to an understanding of the situation and the reasons behind choices made by those who had included backpacking as part of their touristic experience. The vast majority of those interviewed both as volunteers/backpackers were female. This reflected the composition of each of the three groups.

The opportunity to return on three separate occasions to the same place is a major strength of the study because this allowed access to a variety of different situations, especially as the dynamics of the group altered each time depending on who was staying at the volunteer house. This allowed some degree of verification of results to be undertaken as similar issues emerged on each occasion, but also some interesting differences, which will be explored more fully later.

The role of the researcher is also an interesting one, especially within the context of the volunteer situation. Despite a significant age difference, the researcher was accepted into the group because all the volunteers were supportive of each other. This enabled the researcher, for example, to travel to Zanzibar when several volunteers decided to go backpacking during their stay and on another occasion to join another group who went trekking in the mountains of northern Tanzania. So, whilst it was possible to maintain some distance and detachment, it was also possible to gain easy access to the group.

## Discussion

The findings from the research are discussed in detail below. Firstly, I consider the profile and motivations of the volunteer/backpackers I interviewed in Tanzania, followed by a discussion of their backpacker travel experiences before, after and during their volunteering.

The profile of the volunteer tourists observed showed that they were largely composed of young people under the age of 30 and predominantly those who are taking a year 'out' before starting university or between university and starting their first full-time employment. They were, therefore, mainly gap year students. Whilst there were older people who were also volunteering, it seemed to be unusual for this group to also fall within the volunteer/backpacking model as they tended to arrive in Tanzania to volunteer and return home on completion of their project. Most of this group did journey to Zanzibar, but they were more likely to have pre-planned this and to have booked into more expensive accommodation on the island, looking forward to some comfort and luxury. Their trip also tended to be planned at the end of their placement rather than during it.

The greatest numbers of those involved in this study were British, firstly because they were in the majority at the volunteer house and, secondly, they were the group who were more likely to backpack either prior to or after their volunteer placement. Virtually all those at the volunteer house did, however, travel to Zanzibar during their stay irrespective of which country they came from.

Virtually all the volunteer/backpackers had significant travel experience prior to their placement and for several of them this had been backpacking. However, *none* of those interviewed had been to Africa before and those that had been backpacking had done so in Australia, and in once case South America. This certainly had a significant impact on their decision to do 'volunteer tourism' as part of their travel experience in Africa.

Claire (26, British) describes how she was excited before her trip, but as it grew closer she became more apprehensive as she had organised her journey so that she could backpack in Africa for a month before arriving at the volunteer house. She reflected on how the volunteer project offered her a 'comfort side', but for her this was more that she enjoyed the routine of being at the project, which contrasted with the more random experiences of her travels.

For others, the volunteer house and the support of the organisation was critical in their decision-making process. They were nervous about travelling in Africa and had made a decision, therefore, to undertake a volunteer project as a pre-cursor to their backpacking travels.

> Coming to Africa I was more apprehensive than I have been in any other country and just from stories you hear and the kind of vibe you get back in the UK of it being repressive and dangerous but I did feel that because I was doing it with a company I knew I was going to be looked after. (Cara 28, British)

This was particularly evident with some of the women and although they had reflected on the difficulties they may face, they had made a decision that, irrespective of any problems, the opportunity for travel and adventure far out-weighed any misgivings that they may have had.

Hannah (19, British) was not worried about travelling solo, but considered that her lack of language skills could be a disadvantage. Spending three months in the volunteer house and meeting local people on a daily basis, both through her placement and through interactions in her leisure time, had enabled her to learn some Kiswahili. She also felt that by staying for an extended stay in the area, she had started to become more 'streetwise' and looking back on her subsequent back-packing, she felt that she had benefitted enormously from her initial stay at the volunteer project.

One volunteer/backpacker, Antonia, described how she had felt quite uncomfortable travelling as a lone female. She had started her travels in Africa with a volunteer placement in Kenya and after three months she had started to backpack through Kenya and into northern Tanzania. She had 'mentally prepared' herself for her lone travels, but found the attention she received quite challenging and she felt a certain degree of relief when she was able to, once again, live in the more sheltered environment of the volunteer house in Tanzania. She also highlighted a slight feeling of loneliness as a backpacker compared with the more social setting of the volunteer house.

A number of the volunteer/backpackers recognised that their journey to Africa was a unique experience and one that was expensive and involved a long and often difficult journey. A number of the volunteers, for example, had travelled from the USA and others from Australia. For some, it had necessitated working at home for a considerable amount of time, quite often in jobs that they had not particularly enjoyed in order to fund their trip. They had, as a result of this decided to combine both a volunteer opportunity with extended travel on the continent.

> Once I started researching the countries to decide where I wanted to volunteer, I decided I wanted to see more of Africa and since the flight over was going to cost me so much anyway and since I would be going all the way there, it made no sense to just leave afterwards. (Claire 26, British)

For most, it had been the volunteering aspect that had been the principal driver in their decision to go to Africa and this was linked to motivations such as wanting to help and of doing something more worthwhile, echoing the findings of Simpson (2004a). It must be remembered that volunteering combines elements of altruism with perceived benefits to the individual, i.e. self-interest. Rob (19, British) acknowledged that he wanted to do something that had a 'purpose' instead of 'just going and getting drunk and just going out and partying in different parts of the world'. Several volunteers also acknowledged that volunteering would increase their prospects of finding employment on their return home, setting them apart from other applicants. Imogen (22, British), who was about to begin her PGCE on her return to England, definitely saw an advantage in her volunteer work and had, as a result, combined this with an extended journey through Africa. During her travels in South America, Antonia had wanted to volunteer as part of her backpacking, but had been restricted by her partner who did not share her enthusiasm. She had therefore booked a volunteer/backpacking journey through Africa and believed that combining the two experiences had enhanced her pleasure and she had since recommended this type of travel to others.

For some, the opportunity to remain for part of their visit in one place for several weeks was critical in their decision. There were a number of reasons cited for this. Firstly, some saw this as an opportunity to become more involved in the local community, thereby establishing closer links with local people and begin to understand the culture of the people a little more, as suggested in the research carried out by McIntosh and Zahra (2006). Reflecting back on both experiences, volunteering and backpacking, Hannah (19, British) acknowledges that there was 'nothing like volunteering' as even when she was backpacking she still felt like a tourist, whereas over the three months she spent on her placement she felt she had been able to immerse herself in the culture and had managed to make friends with a number of local people. The majority of those interviewed had been invited to the homes of local people and had found this a particularly important part of their trip. While backpackers consider that they would enjoy more contact with locals, there is often a contradiction between what is wanted and what happens in reality (Wilson & Richards, 2008), whereas the volunteer part of the travel experience is considered to offer greater opportunities for contact to be made.

For others, the opportunity to spend several weeks in a volunteer house afforded the opportunity to establish social contacts with like-minded people and unlike the research on backpacking where these relationships are described as often transitory (Cohen, 2004a: 53), many of the volunteers established contacts and friendships that have extended beyond the project. Indeed, the advent of sites such as Facebook has aided the maintenance of these contacts.

Some of the volunteer/backpackers enjoyed the feeling of 'family' afforded by their stay in the house. They describe leaving the house to continue on their travels as 'a horrible experience'. Imogen (22, British) had made some close friends who were Australian and she felt sad at leaving them even though she was moving on to the backpacking part of her trip, which she was excited about. She describes how, when she reached Nairobi on her own, she would willingly have returned to the house 'which felt like home', although subsequently she had had a tremendous time on her travels through Africa.

Hannah also talks of the house becoming 'home':

> I loved every second of being in a house as there was always someone there you could talk to about your day and just chat with. The bedrooms were in dorm style and I loved it even though I had rats living in my room for a few weeks. It was all part of the experience and I wouldn't change it for the world. Our cook and cleaner were amazing. Mama Koleta (the housekeeper) was another mum to everyone even though she spoke next to no English. I still keep in touch with volunteers I met. (Hannah 19, British)

The transition point from backpacking to volunteer and then from volunteer to backpacker is significant. The entry point is characterised by a need to be accepted and to make friends quickly. The arrival is described by several as quite a nerve-racking time as the established community tends to be very tight-knit. Strong friendships are established early on and much excitement is created by new volunteers arriving. Antonia (British, 24) described it as 'rather like the Big Brother house'.

She was particularly upset when several good friends moved on from the house, describing it as rather like a 'grieving' process. Towards the end of their projects, several described how they were ready to move on as they had not replaced their initial strong friendship groups and had begun to feel a detachment from newly arriving volunteers. Antonia described how she felt towards the end, that the house was becoming dominated by strong characters and she had felt ready to continue her backpacking to South Africa.

Imogen (22, British) clearly articulated the reason behind her decision to combine both volunteering and backpacking. She decided to start by volunteering in the Kilimanjaro region because she believed it was an opportunity for her to 'see what Africa was all about'. She had concerns about travelling alone to Africa and had been pleased to meet a fellow volunteer prior to departure at Heathrow, but she also had a desire to satisfy what she saw as a need within herself to travel. This she felt would be achieved, at least for the foreseeable future, by exploring the southern part of Africa after her placement was finished. This sense of a need for adventure was also expressed by all of those interviewed. Each saw themselves as an adventurous person who had a curiosity to 'see' the world.

Whilst Claire (26, British) also had a desire for adventure, she expressed a different reason to volunteer and then to backpack, firstly through Africa, but then to continue around the world on a trip that would last in excess of a year. She described what she considered to be a mid-life crisis and this had led her to want to 'do some good in the world'. She, along with the majority of volunteers, was perhaps in search of what Cohen (2004a: 54) would term the 'constitution or re-constitution of (their) sense of identity', which at a later stage could impact on 'their attitude to their society, and their choices regarding studies, occupations and sexual relations that they make when they return home' (Cohen, 2004a: 54).

Volunteer tourism is unusual as a vacation choice as it has at its centre an intention to work. Vacations and work have in the past been considered as opposites and Graburn suggests that our 'conception of tourism is that it is *not work*' (Graburn, 1989: 22, author's italics). Tourism is thus 'defined by the tendency to spend money rather than to work for it during the trip' (Uriely & Reichel, 2000). Simpson (2004b) suggests that for gap year students there is a desire to distance themselves from the idea that their experience is a holiday and that they try to ensure that it cannot be conceived as one because the experience is difficult, undesirable and unpleasant. This, it is suggested, contradicts the 'normal' experience that is sought for in a holiday. She states that participants she interviewed were particularly keen to embrace material deprivation, such as lack of hairdryers and normal toilets and food. In this context, several of the volunteers commented on the accommodation in a favourable light and expressed surprise that it was as good as it was. They did not feel that a lower standard of accommodation, however, would have put them off or been less desirable.

> I'm surprised at our accommodation. I thought it would be rougher than this. I was really expecting freezing cold showers-possibly no showers and of course not like flushing toilets or anything like that so yeah the accommodation has been surprising. (Imogen 22, British)

Volunteers are characterised as devoting some or all of their leisure time to work activities. However, this view caused controversy amongst the volunteers, some of whom considered that volunteer tourism was definitely a work activity, whereas others considered that it was more like a holiday, as this extract demonstrates:

> **Kate (Australian, 22):** Yeah coming out I definitely thought it was going to be more work but actually having the experience of being here definitely more holiday.
> **Kath:** So you feel there's pretty much a push towards you being a tourist.
> **Kate:** I feel pretty much that the whole programme it's pretty much a holiday. You do safaris and you do some volunteering on the side. That's the way I'm feeling at the moment. It's primarily holidays and safaris and if you can do some work on the side that's... yeah...
> **Kath:** You give me the impression you would have liked to come here and really rolled your sleeves up and got immersed in it.
> **Kate:** Yeah and that's what I expected the reality to be for it to be- like that.

The volunteer house can, therefore, be characterised as an enclave to which it is possible to retreat, with what Wilson and Richards (2008) describe as 'cultural work' being undertaken outside of it. In the case of the volunteer/backpacker, this 'cultural work' is arguably more intense than that of the backpacker, as it involves periods of deep interaction with local people especially during those times spent on the various projects. Whether this can be viewed as 'real' work is difficult to ascertain. Perceptions vary from person to person and are not always dependent on the length of stay or the number of hours spent on a particular project. Most volunteers were unable to articulate their feelings about this, although some viewed it definitely as work, whilst others saw their experience as more of a holiday. What is certain is that there was a need for the retreat to the 'metaworld of the enclave' (Hottola, 2005).

Importantly, however, for many of the volunteers there was a need to have a 'holiday' from the 'work' involved in the volunteer project, especially as a significant number of those who had volunteered had done so for periods of between six and twelve weeks. This led many of the volunteers to engage in backpacking trips *during* the volunteer placement.

The travel brochure and web-site of the company emphasise the attractions of the northern part of Tanzania with a backdrop of Kilimanjaro, the wide-open spaces of the Serengeti and particularly, in the context of backpacking, the island of Zanzibar. The majority of younger people who undertook volunteer placements in the Kilimanjaro region also travelled for up to a week, first to Dar es Salaam and then via the ferry to the island, spending a number of nights on the beaches to the north and east of the island and then visiting Stone Town before returning to the volunteer house. Most people travelled in small groups, largely formed whilst in the house. Planning took place using the *Lonely Planet* guidebook, but accommodation was largely chosen through word-of-mouth from other volunteers who had just returned. Whilst virtually all the volunteers travelled to the island, it was significant that it was the younger volunteers who sought out budget-price accommodation. Older volunteers were also much more likely to plan their trip to Zanzibar *after* their volunteering was completed, whereas younger volunteers were more likely to travel *during* their placement, and were influenced much more by when their new friends were intending to travel. This was definitely a holiday time with opportunities taken to party and drink excessively, to sunbathe, to swim and to snorkel. The volunteers/ backpackers were keen to enjoy a hedonistic lifestyle for a brief period. Whilst social activities and heavy drinking were also the norm on several occasions *within* the volunteer house, depending on the mix of volunteers at particular times, this was set very much in the context of a realisation that there was 'work' to attend to the next day and that a hangover was not an excuse for missing the volunteer project.

There were a number of very different reasons given for this travel during the volunteering and the switch from volunteer to backpacker. The first was simply that after, for example, six weeks of work teaching English and maths in a nursery or secondary school, the volunteer needed a rest to 'recharge his/her batteries'. There was, in fact, a need for a 'holiday'. This equated very much with the typical idea of a holiday at home. Packing was done, plans were made, passports safely stored, currency obtained, their new 'family' waved them off and welcomed the travellers back after their break, when tales were told of the experiences of the island in much the same way as would happen in a holiday at home. This was also combined with the desire to maximise opportunities in what could be for many, a 'once-in-a-lifetime' experience.

> I think it can be a bit stressful and intense to do three months with no break in a country where they don't speak our language. When I came back I felt revived and had a burst of energy towards the project especially in terms of the lessons I did – I was more inventive. The time in Zanzibar had given me time to think about what lessons etc I could do to improve their English/maths, as before I got there and was planning lessons I had no real understanding of their capabilities. (Jodie 19, British)

Secondly, an important reason for volunteers' decisions to travel whilst on their project was that, while many had come with a real desire to help and 'to make a difference' albeit in a very small way, they soon realised that this was difficult to achieve. Projects were often poorly organised and several volunteers had placements where they were not needed at all or where there were substantial amounts of time when the school was on holiday or exams were taking place. James (29, British), who had travelled on a motorbike from England, did not have any experience of volunteering during his stay in the house as there was no project available for him. He had wanted to teach football in the area, but this could not be arranged and his school placement was during a school holiday. Kevin (21, British) also had a similar experience and had had to find his own placement as he too wanted to teach football, but had been allocated a teaching project in a school, which he did not feel comfortable doing. Whilst these were exceptional cases, a significant number of volunteers were frustrated and disappointed by their projects and felt that they were not really needed. Two Australians, Kate and Jo, reflected on their placement in a women's centre, with Kate voicing their thoughts thus:

> I feel disappointed because I feel I'm not very hands on and it's not sort of nine to five. I thought it'd just be working, working, working and that would be the focus but I've just like been teaching two hours a day and it's just like two students so I feel like I'm just wasting my time when I could be doing things to help. (Kate 22, Australian)

For some this resulted in them undertaking trips whilst volunteering, which they possibly would not have done if the experience had been more challenging and perceived as more important. Rhiannon (22, Australian), who was working in an orphanage and did feel that she was doing something valuable, was reluctant to go on trips other than at week-ends and at the end of her project. She felt that she would be letting people down and that there was a genuine need for her work. She had a long, hard day, often not arriving back at the house until dusk.

Thirdly, there was a lack of consistency shown by the volunteer organisers who were running the house where the volunteers lived. When trips, especially safaris, were organised by the in-country team, it was considered acceptable and desirable to take time away from the project. However, when these trips were of an independent backpacking nature, this was considered inappropriate and efforts were made to prevent this happening. There were also concerns over individuals' safety, which was seen as very important, but this was often not appreciated by young travellers who were keen to explore and to undertake adventures whilst away from home. This often led to confrontation and on a number of occasions threats were made regarding withdrawal of visas and possible deportation from the country. Despite this, a significant number of volunteers undertook backpacking trips during their placement.

## Conclusion

As the tourism market continues to explore new avenues to entice the traveller, the concept of volunteering as part of the holiday has become an increasingly attractive option, especially for younger people. This largely, but not exclusively, gap year market has made up a significant proportion of those who have chosen backpacking as a way of gaining an ever-increasing number of new and exciting experiences. Volunteering has also grown in popularity and the opportunity to combine both backpacking and voluntary work in Third World destinations is becoming increasingly common. Volunteer tourism, which offers the young traveller an institutionalised experience, is often contrasted with the more flexible and less constrained backpacking travel. Volunteer houses such as the one in the Kilimanjaro region of Tanzania provide a suitable enclavic space from which to explore and experience the culture of the area in a way that is often more intense than that of the average backpacker, but with the safety net of a travel organisation providing reassurance and security.

For these volunteers, unlike backpackers, the 'work' outside the enclave is not merely an encounter with the 'other', but is in reality, at least to some extent, 'real work' as it is defined in our Western culture, although usually without any financial gain. Despite this lack of payment, volunteers, it has been argued, take part in such activities with an expectation of some self-gain, whether to enhance employment prospects, to be able to access the local culture more easily or simply to spend time with like-minded people and to undertake a trip that, as Hannah reflects, she would do again 'in a heartbeat'. This is combined with a degree of altruism, although volunteers do not readily recognise this as a motivation. What is certain is that those who have combined their backpacking with some volunteering believe that this experience has added to their enjoyment and provided memories that they continue to share with friends they have made during their placement. In this chapter, the experiences of volunteer/backpackers has been explored and some understanding of their reasons for making such a choice. The idea of volunteer tourism and the dualism of work and leisure, has led to a consideration of a 'holiday within a holiday', where backpacking activities have taken place during the volunteer placement. Such mobilities set within the backpacking experience afford key areas for further research as the demand for multi-experience travel increases.

## Chapter 11

# Backpackers in Norway: Landscapes, Ties and Platforms

GARETH BUTLER

## Introduction

Hannam and Ateljevic (2007b) have acknowledged the rising importance of backpacker tourism from both a social and economic perspective, and this mode of travel is now seen as a highly dynamic and quickly evolving market segment. Despite these revelations, previous research has been heavily swayed towards an empirical or quantitative approach and, as a consequence, has neglected other 'deeper' aspects of this mode of travel (Hannam & Ateljevic, 2007b: 18). Moreover, backpacker tourism research does not encompass a wider range of geographical destinations, leaving a notable void of research for particular regions. Indeed, Wilson *et al.* (2007) have acknowledged the increasing depth of research on 'budget' and 'youth' travel, but concur that an apparent void still exists in the context of European-based research. Effectively, Europe is identified as a *source* of outbound backpacker travellers and not as a destination per se (Hannam & Ateljevic, 2007b). Furthermore, Wilson *et al.* (2007: 195–196) express that an added detrimental effect of this lack of research on Europe has meant that non-European backpackers, such as North Americans, South Africans and even South Americans, have also been neglected as recognised participants of backpacker research. Citing the 'OE', a common term used for European trips by Australasians and New Zealanders, Wilson *et al.* (2007: 186–187) elaborate on the variety and breadth of what Europe has to offer as a backpacker destination. During their stays, Australians and New Zealanders could be found indulging in the culture of Pamplona, Barcelona and Rome, attending the Oktoberfest in Munich and island hopping across the myriad Greek Islands.

Backpacker hubs or enclaves in South-East Asia and Australasia are significantly populated by European travellers, but equivalent destinations in Europe that have been experiencing high volumes of non-Europeans, have been overlooked in comparison. Ateljevic and Doorne (2007: 66) identify this problem, citing the research of Shipway (2000) that investigated backpacking in Europe as opposed to Australasia, as a 'rare exception'. Wilson and Richards' (2007: 23) research additionally discovered that current examinations of backpacker travel still centre upon more traditional or 'exotic' locations, whereby studies were largely found to focus upon popular destinations such as South-East Asia or India and, as a consequence, has limited the research conducted into the 'backpacker experience'.

A consequence of this has been a lack of awareness in understanding the economic potential and the social characteristics of participants in the European market. More specifically, little research has been conducted in Norway with a specific goal of monitoring and assessing the diversifying trends and complex profiles of backpackers. Hence, this chapter will attempt to shed light on the motivations of backpacking hostel users choosing Norway as a destination for travel.

## The 'Erosion' of the Typical Backpacker

Contemporary research into backpacker identification suggests that they are from a small range of western countries, ethnically white, university- or college-educated and of a middle class upbringing (O'Reilly, 2006; Moshin & Ryan, 2003). However, recent research in the field has begun to challenge these criteria and reveals that such narrow demographic profiles are no longer accurate. Although these characteristics still represent the majority of backpackers, it has since been argued that this sector is becoming more multifaceted than ever. Maoz (2007) cited the emergence of Israelis, Japanese and other Asian nationalities as evidence to indicate an 'erosion' of the contention that backpacking is a predominantly European, North American and Australasian activity (see also Westerhausen, 2002). Effectively, the scope of participants in backpacker travel is now too wide to adequately categorise; perhaps Sørensen (2003: 852) best summarises the problem of defining this complicated mode of travel: 'The variation and fractionation make it all but impossible to subsume all the above-mentioned individuals and groupings under one uniform category, for it would be so broad as to be devoid of significance'. The emergence of a variety of backpacker subcategories further underlines the increasing difficulties surrounding backpacker definitions, and Sørensen (2003: 852) highlights that the backpacker in a modern context is 'a social constructed identity rather than a clearly defined category'.

It is widely believed that backpacking is more associated with 'self-definition' as opposed to 'conformity to a set description' and the vast majority of individuals would reveal many demographic, characteristic or motivational differences from the next. It appears that the classification of the modern backpacker would be a futile attempt, particularly attempting to do so using a set of demographic criteria. To clearly understand the concept of backpacking, it must be remembered that the originally imagined profiles of backpackers are no longer necessarily applicable due to the acknowledgement that a wider range of people from all backgrounds are now partaking in this activity. The days of Cohen's (1973) 'aimless drifter' is now an ever-increasing rarity, particularly as the mainstream begin to engage with the idea of 'backpacking'. Cohen (2004a: 50) ponders the question as to whether backpackers are now merely the 'trend-setters' for postmodern tourism, who simply lay the tracks for more conventional tourists, who then further dilute the perceived typology of the conventional backpacker. Additional research suggests that the profiles from both an economic and socio-cultural perspective are diversifying, as well as the general demographic expansion of participants. The consequence of this realisation is that future definitions will need to be more pro-active and responsive to change and that businesses and accommodation sectors who wish to attract their custom, must also be aware of these sensitive changes. These definitions should encompass a series of subgroups to cater for the various niches that exist, while at the same time disabling the need for a general broader categorisation process. Ateljevic and Doorne (2006: 61) concur with these views, suggesting that any attempts to understand the concepts and characteristics of backpacking should acknowledge that a constant process of re-definition must take place to understand a dynamic and evolving market segment.

## Backpacker Tourism Mobilities

According to Hall and Page (2009: 6), one of the most interesting aspects of current tourism research has been the incorporation of mobilities and its intertwined relationship. Larsen *et al.* (2007: 249) explain that due to the increased mobilities of postmodern society, tourism has now been seen to be incorporated into the lives of a variety of individuals and groups, such as business personnel, second home-owners and their respective friends and families, exchange students and gap-year workers abroad, migrants and former refugees, and people with distant friends and families. Larsen (2001: 81) comments that: 'Modern tourism is a reflection of, and indeed constitutive of modernity's mobility; tourism by definition involves geographical performances of corporeal mobility through physical space via mobility technologies or vehicles'. Furthermore, Galani-Moutafi (2000: 220) explains that the studies of mobility have enabled the deconstruction of previously rigid and 'ethnocentric categories', which have frequently been formulated in an attempt to understand tourism processes and distinguish clear differences, such as the 'self' and 'other' or the 'familiar' and the 'exotic'. For many, travel is now seen to be too simplistic, and the potential adventures of many travellers have been sanitised, while at the same time the activities on offer at the destination have now been tailored, synthesised or pre-packaged ready for consumption on arrival, which offer instant 'pleasure-inducing' impacts, a product of postmodern culture according to Galani-Moutafi (2000: 211). For the more adventurous travellers or Urry's (2002b) 'post-tourist', this is the very antithesis of the travel concept and escapism from such unwanted scenarios is now becoming a challenge in its own right. A large number of tourists have been empowered to go further and to experience new destinations, but paradoxically this has resulted in the original long-distance travellers losing their relative freedom as their choices of destination become narrower as they are constantly on the move to escape the rest (Westerhausen, 2002).

According to Jacobsen (2004: 6), holiday mobility is an 'essential feature of contemporary European life' and this can be witnessed by the mass exodus of motorists to areas of natural beauty. The Scandinavian region appears to be one of the key destinations for such tourists who are keen to experience 'untouched nature' and 'unique sights' (Jacobsen, 2001: 102). Among the main benefits of travel via car in Europe, are the opportunities to experience 'virtual otherness' (Larsen, 2001: 81), whereby the tourist can encounter vistas and landscapes en route. Jacobsen (2001: 108) suggests that this 'sightseeing at a swift pace' provides a range of highly sought sensations of places to the traveller in a relatively brief passage of time and argues that this phenomenon relates closely to the activities of the modern backpacker. Jacobsen (2001: 108) further suggests that for both groups 'being on the go' is perhaps the most important aspect of the journey.

However, despite these assertions regarding the liberation of the traveller, Larsen (2001: 88) is critical of the involvement of the car in the modern travelling experience, and terms modern vehicles as 'inhuman machine hybrids of mobility and visualisation'. Although the car affords the driver and the passenger a great deal of extra freedom, there are constraints to be faced. Larsen (2001: 89) highlights these drawbacks as follows:

> The passenger is a spectator that can only sense the "outside" world through the visual sense; the sounds, tastes, temperatures and smells of the countryside are to a large degree "reduced" to a framed, horizontal visionscape... The train and the car provide a hypersensuous experience of seeing which deliberately comes to dominate all the other senses. It is this sense that we can clearly understand these mobility machines as simultaneous vision machines.

Furthermore, Jacobsen (2004: 9) concedes that the escapism offered via mobile tourism through landscapes and scenery may just be as equally important as the destination they visit.

Dann (1999) and Buzzard (1993) suggest that tourists are constantly seeking change and are simultaneously attempting to avoid the routines and obligations they face every day at home. Brown (2007: 364) identifies that the action of travel requires a variety of 'practical organizational activities', which are considered to be part of the 'mundane' processes associated with preparation before the journey takes place. These processes, such as arranging transport, finding accommodation, shopping for sun tan lotions and beachwear, for example, are seen as unavoidable criteria that Trauer and Ryan (2005: 486) identify as 'ritualistic behaviours' and are necessary before they depart for their 'temporal escape from the ordinary'.

While these rituals are deemed to be part and parcel of the travel planning process for many, these practices are not merely superficial activities, but highly detailed arrangements that must be negotiated by the participant. In addition, it could also be argued that these rituals and mundane encounters do not cease on embarkation of the journey, but systematically continue throughout the journey. According to Brown (2007: 369), tourists face four general 'mundane problems': tourists need to decide *what* activities to do, *how* to do those activities, *when* to do them and, finally, *where* those activities are (and how they can get there).

Brown (2007: 377–378) additionally cites Cheong and Miller (2000) who argue that these problems have been largely negotiated by travel agents and tour guides, who not only solve these problems, but also control and restrict the movements, behaviours and thoughts of tourists who consult them. Despite the negative perceptions associated with such occurrences, it has been counter-argued that many tourists require such dictation as to their vacational experiences and indeed, unwittingly seek 'mundane' experiences. Furthermore, Hannam and Ateljevic (2007b) highlight the institutionalisation and standardisation of backpacking as a mode of travel and it has often been misinterpreted as a negative outcome by many academics. Additionally, it can be argued that such processes are responsive to the demands by travellers who seek the adventure and excitement of a new destination, yet simultaneously require familiar surroundings, such as chain hostels and backpacker-themed bars. Hannam and Ateljevic (2007b) have identified the back-packer enclave as a particular example of how destinations associated with perceived liberal and free-thinking travellers are now paradoxically areas of familiarity and standardisation – the original antithesis of the backpackers' travel agenda. Enclaves are effectively zones of mediation, where participants can encounter 'suspended' extremes, such as famil-iarity and difference in the same context (Wilson & Richards, 2007: 22). Traditional enclaves, such as the immediate area around the Khao San Road in Bangkok, or the Kings Cross suburb of Sydney, are excellent examples of these suspended extremes where backpackers can be found congregating en masse. At one end of the spectrum, the backpacker has a range of familiar products at their disposal; the hostel, traveller-themed bars, internet cafés and backpacker-orientated tour operators. At the other end of the spectrum, the backpacker has a 'world of difference' at their disposal whereby the participant can experience local culture via restaurants, nearby attractions and other points of interest without becoming too deeply submerged into the 'unknown'. Andersson Cederholm (2004: 226) explains the standardisation of the backpacking mode of travel:

> The notion of contrast and liminality was socially constructed within the structural conditions of backpacking as a gradually institutiona-lized phenomenon. Young westerners of today, who go for a rite of passage-like journey for a few months before they settle down with anticipated adult responsibilities... find an infrastructure of travel agencies, transportation, hostels and restaurants with backpackers as prime customers, as well as media like travel magazines, guidebooks and novels about backpacking.

Although such outcomes are naturally perceived to be negative by-products of the growth of backpacking, there appears to be a fine line between attraction and rejection of such facilities. Citing MacKay and Fesenmaier (1997: 542), Prentice (2004: 925) argues that although over-familiarity prompts a desire to reject standardised tourism products, familiarity in itself is an attractive to feature to many tourists and travellers even if the latter might deny such an admission. Nordstrom

(2004: 61) additionally argues that tourists revisit destinations that they like because it reduces the risk of uncertainty associated with holiday-making due to familiarity. The backpacker enclave appears to dispel the notion that the sole consumer of familiarity, standardisation and institutionalised products is the common mass tourist. According to Gibson and Yiannakis (2002), cited in Hyde (2008: 716), the desire for familiarity and institutionalised facilities increases with age, as does the lesser requirement for 'novelty-seeking', but other factors may also play a role in whether such approaches are adopted. For example, Uriely *et al.* (2002: 521) suggest that the avoidance of conventional facilities may be attributed to budget limitations, and that they travel as backpackers not because they attempt to discover 'meaning' or because they hold 'anti-establishment views', but because they cannot afford to stay in more comfortable surroundings. Citing Cohen (1973), Uriely *et al.* (2002: 523) further argue that backpackers are additionally often found to be 'inward-orientated' and, effectively, participants of a prolonged summer trip. These backpackers fail to interact with locals, and establish ties and communications only with those who exhibit similar demographic characteristics to themselves. Effectively, it appears that many back-packers have the potential to exhibit similar characteristics to mass tourists by travelling in 'environmental bubbles' (Uriely *et al.*, 2002: 522), whereby the tourist is in a simultaneously 'foreign' and 'familiar' place. Although the backpacker or tourist would loath to admit their desires for familiarity and tastes of home, the rise of the backpacker enclave suggest that they are highly frequent participants of such activities.

## Methodology

The research, on which this chapter is based, was completed between late April and August 2008 and focused upon acquiring an extensive array of qualitative data from a number of guests at hostels in Norway. In the interests of logistical practicality, it was decided that the research focus region would centre on southern and western Norway and that no hostels would be visited further north than Trondheim. A sample size of around 20–25 hostels was deemed to be an accurate enough representa-tion (internet search estimates suggest around 90–100 exist throughout the country). To determine which hostels would be selected, the two most recommended 'classic routes' from *Lonely Planet's Norway* guide book were selected as the basis. The first itinerary, which was titled 'Norway in a microcosm', involved a combination of some of Norway's largest cities such as Oslo, Bergen and Stavanger, popular tourist towns such as Flåm and Voss and several of its naturally beautiful regions such as Lysefjord and Hardangerfjord, to give a largely contrasting and variety of experiences. The second itinerary, 'The Heart of Norway and the best of the Fjords', focuses more upon the geographical beauty of Norway, but at the same time incorporates particular tourism hotspots such as Lillehammer and Ålesund. Using these two routes as a 'rough guide', both itineraries were then plotted onto a road map. All of the major locations recommended by *Lonely Planet* were identified and a logical route to reach them was constructed using the most likely major roads or typical route recommendations. From this stage, all hostels that were passed via either route were then selected as hostels that would be used for the research. Both unstructured guest interviews and ethnographic diary accounts were compiled and the preliminary findings are revealed in this chapter.

## Why Norway?

Participants were asked to discuss why they chose Norway and the motivational factors behind their decisions. Three distinct categories of respondent appeared to emerge; the highly motivated destination visitor, the obligated visitor and the opportunistic visitor. The highly motivated destination visitor was someone who had chosen to visit Norway for a highly specific or clearly defined reason, while the obligated visitor was someone who had arrived in Norway as part of an obligation, namely work or for study purposes. The third group, the opportunistic visitor, was a guest who had chosen to visit Norway for external, non-destination-related reasons, such as a cheap break 'somewhere' or an opportunity to meet up with old friends for example. In the latter scenario, the destination was a secondary factor in the holiday decision-making process, where the most important factor was 'getting away', wherever that may have been. In what follows, I have isolated three particular motivations that backpackers to Norway expressed consistently, firstly, the importance of the Norwegian landscape, secondly, the importance of having historical familial ties to Norway; and thirdly, the use of Norway as a 'platform' for particular encounters.

### Landscapes

For the highly motivated backpackers, such as Sung, from South Korea, the motivations were highly destination-specific, concentrating mostly on specific aspects of the Norwegian *landscape*: 'To see the fjords, countryside, nature and fresh air. Just something which is very different from Seoul'. Others, such as Jeroen from the Netherlands, revealed that the decision to visit Norway was a highly motivated desire held for a number of years:

> Its always been a lifelong ambition to go to the North Cape on my motorbike. I just always had this dream of riding through the mountains and fjords and being totally free from everything back home. I've waited five years for this trip and its going to take me nearly three months to complete it all. It has taken a long time for me to be able to get this amoung of time off from work, but so far its been worth the wait. (Jeroen, Netherlands)

Michael from Germany also exhibited a similar range of motivations for his visit to Norway and returned on the basis of a previous experience:

> Its always been been one of my favourite places. I came here a few years ago on a tour to the North Cape and was hooked. Everything is just so big and the roads are great for driving, not like in Germany with the jams. Here I'm alone or at least I feel like it. It's a really special feeling being on the road without anyone around. Just you and nature. (Michael, Germany)

Daniel also from Germany, suggested that 'it was always one of the places that I wanted to visit', and added that 'it was the one place in Europe I really wanted to go because of the landscapes'. For Hanne, a Danish visitor travelling with two female companions, Norway offered important criteria, such as 'familiarity' and 'safety', while simultaneously being an 'incredible place' with 'amazing scenery'.

The theme of scenery appeared to be a common motivational factor amongst those highly motivated to visit Norway. With the exception of Sung from South Korea and one other interviewee, all who expressed a desire to view the vistas of Norway travelled throughout the country via either their own cars or motorcycles and revealed a high degree of vacational mobility (Jacobsen, 2004). The requirement to experience the mountains, fjords and overall wilderness appeared to coincide with Jacobsen's (2001) notion of sightseeing at a swift pace in the vast majority of cases. Even those on long distance/duration journeys, such as Jeroen and Michael, felt compelled to travel as 'far as possible' each day, in order to maximise their time and see as much as they could. In these instances, the destination at the end of each day was merely a place to rest as opposed to a nodal point along a carefully constructed touring itinerary. Jeroen admitted that his stay in Trondheim was merely coincidental and that his stay was influenced due to rising fatigue. Michael's stop at Stavanger again was motivated by respite as opposed to the city itself and he conceded that he would spend little time exploring during his stay at the local hostel. In several scenarios, cities and towns acted as unplanned places of rest between lengthy road journeys through Norway's landscapes, and with the exception of several travellers interviewed at Bergen, few revealed any motivation for choosing the actual places where they stopped. In line with the views of Jacobsen (2004), it appeared that the act of moving throughout these landscapes superceded the desires of visiting or more accurately 'stopping' at particular places. 'Stopping' was seen to be a literal postponement of the journey and was only found to occur when the traveller deemed it necessary (primarily for sustenance or sleep). Only Hanne and her friends from Denmark revealed a greater desire to visit places as opposed to the motorised journeys between them, and for her, the landscapes on offer were a 'bonus' while in transit from one place to the next.

### Ties

For two separate American visitors, the decision to go to Norway was more of a personal journey of self-discovery as opposed to a visually or environmentally pleasing one. The search for heritage or family roots was the particular motive of these two respondents, as both sought to see the home of one or both of their parents. Yeoman *et al.* (2007: 1135) have suggested that many tourists are on the trail of authenticity and seeking a 'connection' with a destination, which hypothetically provides both roots and something that is perceived to be 'real'. Karl was on a self-proclaimed 'journey' to establish severed ties with his Norwegian family. Both parents were Norwegian and Karl himself was born and raised in Oslo before his parents divorced and he grew up with his mother and brother in the USA. The journey to Norway appeared to be, simultaneously, an obligation to meet ageing relatives and engage in an adventure in an unknown place that was paradoxically home:

> Well, I was actually born here and raised in the US. My parents are both Norwegian but seperated when I was young. My dad stayed here and I moved to the US with my Mom. I've never been here since I was 3 and I don't remember anything. I guess I wanted to get a sense of what "home" is. Its strange to be from a country you don't know, I guess its like a journey of discovery and to find out more about my family and where they're from. I suppose I just took the

opportunity to take a break and travel but at the same time to learn something about who I am. (Karl, USA)

The reasons for Karl's journey appear to strongly relate to the work of Coles and Timothy (2004), who discussed the increasing importance of the diaspora in social movements. In this instance, it was suggested that greater numbers of people were increasingly on the search for links to a particular identity. In the case of Karl, the journey to Norway was seen as an opportunity to re-establish links to this community, where his family and albeit briefly himself, had originated. Coles and Timothy (2004: 3) reveal the movements of the diaspora and the notion that they are 'drawn together' in an attempt to reaffirm close ties:

Definitions and conceptualizations of diaspora are fluid and con-tested and have been the focus of considerable debate. Diasporas are groups of people scattered across the world but drawn together as a community by their actual (and in some cases perceived or imagined) common bonds of ethnicity, culture, religion, national identity and, sometimes, race.

Although Karl appeared to exhibit a range of highly motivated reasons for his journey, another respondent, Melissa, appeared not completely sure about where she was going or even why she was going. In this instance, it could be argued that she was drawn by the 'perceived' or 'imagined' common bonds, but still failed to adequately justify her expectations:

My mother is from Norway so I decided to finally come and visit a few weeks ago. I'm living in Barcelona for a year now and I'm going back to Chicago soon to study. I've been all over; Prague, London, Paris, Rome...but I guess I had to come here too. It was like now or never. I didn't really know anything about the place so I chose the three largest cities in Norway, and that was the basis for me coming here [to Trondheim]. (Melissa, USA)

Melissa further commented that 'I guess I had to come' and reveals the degree of obligation in her motivations for her arrival in Norway. Her statement relates to the findings of Poria *et al.* (2003: 250), who revealed that many people who are in search of their own backgrounds are at least partly motivated by a 'feeling of obligation'. Additionally, she appears to place pressure on herself by creating a 'now or never' scenario, whereby if the opportunity is missed, she will never be able to go again. In this scenario, the respondent appeared to reveal a lack of understanding about who she was and had consequently questioned her own 'self-identity' (Desforges, 2000). In effect, Melissa felt drawn to Norway via the perception of an imagined community, whereby she would encounter people of the same culture, ethnicity and nationality as her mother. Despite this common ground, however, Melissa still felt isolated and revealed the disappointment of her inability to establish any hidden insights into her journey:

> I just guess I wanted to learn a little bit about my mom, and maybe even me too. However, I don't really feel close to this place. I don't feel any form of belonging and I'm pretty disappointed about that. I thought it would maybe feel like belonging or something, but it just doesn't. I feel nothing. (Melissa, USA)

Thus the search for ties in Norway appeared to result in two very different outcomes for the two travellers. Karl spoke of his rewarding experiences, which consisted of acquiring visual aids and images of what 'home', in the Norwegian sense, was actually like. In addition, Karl also managed to re-establish contacts with cousins and elderly relatives who had not been encountered for over 20 years. This encounter gave further meaning to the images, which Melissa could not attain. As Poria *et al.* (2003: 249) suggest, an attraction or destination is space that allows those in search of heritage, an opportunity in which they can relate to. Although this desire differentiates from those simply in search of the 'gaze', it appears that this does not necessarily guarantee that they will attain any greater enjoyment or meaning from the experience. Although Melissa encountered a similar arrangement of vistas and landscapes to Karl, she failed to give them meaning due to the lack of personal contact with any blood-relations in Norway. As a result, Melissa failed to give any further meaning to the concept of her Norwegian heritage and realised that she had, in fact, little in common with her 'own' people.

## Platforms

The third group, the opportunistic visitor, was a guest who had chosen to visit Norway for external, non-destination-related reasons. Several examples of such visitors were encountered, almost exclusively in either Oslo or Bergen. Here, the motivation for travel was to find a platform where a social encounter could take place or somewhere where mundane lives or jobs could be temporarily left behind. In the case of several interviewees, Norway was chosen as a platform for a variety of different outcomes or 'events' to take place. For those like David, the desire to escape the mundane experiences of work and home were commonly cited reason, and almost anywhere other than home would do:

> For me it was one of the cheapest places to go for a short break. I just went on the Ryanair website and looked for the lowest prices, I didn't really care where. I actually booked the flights before I knew I could get the time off from work. The flights were that cheap that I could afford to lose the tickets if I couldn't go. Oslo has never been that high on my list of priorities, but I thought why not? I just needed a break, anywhere would have probably done... I just wanted to get away in the end. Germany, Italy, France, wherever was cheapest and wherever I had never been before. (David, Spain)

Similarly for Stephen and his former university friends, Norway was a destination of circumstance as opposed to a destination-specific motivation:

> Meeting up was the priority over everything else. We needed to find a time we could all meet up and then we just searched for the cheapest deals for this weekend. It could have easily anwhere else for that matter....We got some cheap flights from Stansted and a chance to do something away from home. It wasn't anything more than that really. It's the first time we've met up since our university days so it was really just a location to meet up rather than picking a specific place. (Stephen, UK)

James and his friends from the UK, were on the search for something a 'little different', but simultaneously were unsure of what that difference should be. As a result, they were undecided on their actual destination until the final moments, before again, cheap travel opportunities presented them with the option of Norway:

> Me and my friends decided to have a long-weekend away. We wanted to do something a little different to the bars and clubs and stuff back home, so we decided to get some cheap flights and go away...just to get away and break the routines of home. We've all just finished our A-levels and been working in part-time jobs. I think we just wanted to reward ourselves but we couldn't really afford two weeks in Ibiza or Magaluf so that's why we ended up here...we'd have gone anywhere to be honest. (James, UK)

All three accounts revealed that the destination was a secondary motivational factor in the planning of the vacation. For Stephen and his friends, the timing of the visit was far more crucial than the specific location of their post-university rendezvous. David also conceded that he would visit practically anywhere as long as it offered a platform to 'get away' and that almost anywhere in Europe would have done. For James and his companions, they use the phrase 'ended up' to signify their lack of control of the actual final destination, which was largely influenced by financial constraints. Here, the crucial factor in their motivation for travel to Norway was breaking the routines of home and a 'reward' for completing studies.

However, the search for platforms was not the only significant factor amongst those with no destination-related motives for travel. Even those who were highly motivated to visit Norway itself, exhibited additional reasons for their choice of travel. Although Michael from Germany had a strong desire to visit Norway on his motorcycle, he additionally revealed that his motives were also governed by a desire to escape home and his 'busy life' and temporarily find 'peace and quiet'. Jeroen, a fellow motorcyclist, also revealed that his ambition to visit Norway was coupled with a desire to experience a 'different world' and be in control of his own destiny through his own vehicle:

> It was very simple [his decision to go to Norway]. Its been an ambition since my teens. Its just a different world in comparison to life in the Netherlands where everything is flat and there are cars everywhere. Everything feels built up, but here is different. This feels like a totally different world where I can ride my bike through some amazing scenery. It's a very unique feeling. (Jeroen, Netherlands)

These journeys offered a sense of freedom and control, where the itinerary could be tailored *ad hoc* during the journey itself. The destination was a key motivational factor to vacation decision-making processes for these interviewees, but equally were the ways in which they undertook their journeys on arrival. For Michael and Jeroen, Norway in itself was a 'favourite place' or an 'ambition', but this alone was not enough. They still needed to be in complete control of how they consumed their journeys and additionally the stimuli they would encounter to fulfil their additional motivational desires for visiting Norway. The destination, by itself, was effectively inadequate even for those with a highly motivated desire to visit the specific place. For Daniel, even a late change of destination had not affected the deeper motivational factors behind the desire to visit Norway:

> I originally planned to go to Tibet [before the political troubles broke out] and then although that wasn't possible I still planned to go to

Sichuan, so I wasn't too worried. Then they had the earthquake, so I had to drastically change my plans. Despite things not going to plan, I feel I can get many of the same experiences I wanted here [in Norway]. I wanted to be alone and to experience travelling some-where remote and away from the cities and suburban life back home. In many ways this has been a blessing, I'm very happy things turned out this way now. (Daniel, Germany)

Although Daniel revealed disappointment at not being able to visit his first choice destinations in China, Norway proved to be a rewarding contingency plan because it offered similar experiential features, such as tranquillity and remoteness. For Daniel, it proved that the destination was not only a conscious choice, but also a platform to deeper social and psychological requirements. Yeoman *et al.* (2007: 1135) suggest that particular environments offer the idea of 'unmediated experiences' and offer the participant the feeling that 'it's just me and the mountain'. In the case of Daniel, the wilderness and landscapes satisfied an 'inner need', which goes far deeper than aesthetic pleasure. For those travelling via their own motor vehicles, time always appeared to be a restraint, and stopping, particularly in urban centres, was seen as wasted time and a delay of the real journey. In line with the views of Yeoman *et al.* (2007: 1132), those in search of landscapes and tranquillity conceded that the only places they would stop by choice were in areas of natural beauty, and that the landscape became a 'time oasis'. Here, journeys could naturally pause and time was no longer recognised as a constraint or obstacle, because the landscape was seen as 'authentic'.

## Conclusion

It appears that users of hostel accommodation in Norway reveal a broad range of differing characteristics and are exemplary of the contention that backpackers, the mainstay of the hostel, are more multifaceted than ever. Visitors of all types, revealing many different profiles were encountered, which further counters the suggestion that contemporary backpackers can be clearly defined based upon demo-graphic generalisations. Visitors ranged from their late teens to beyond retirement age, and several additionally originated from a number of different locations, such as Brazil, Kazakhstan, Japan and Ghana. This discovery appears to concur with the views of Maoz (2007) regarding the emergence of new international sources to the backpacker mode of travel. Simultaneously, backpackers in Norway appeared to reveal distinct motivations for travel and appeared focused as to why they were in Norway. Some visited the region to encounter specific experiences and landscapes, while others came because of obligations such as work or to study. Several respondents suggested that Norway offered a platform to simply 'escape', be it from routines, work or study. Although several conceded that the chosen destination was partly or indeed highly irrelevant to their destination plans, they still revealed a specific reason for their journey: most notably to use the destination as a platform for social encounters, reunions or simply to 'get away', from Buzzard (1993) and Dann's (1999) mundane lives back home. The escape from the mundane world was a commonly cited reason amongst many participants of the study, but simultaneously, few appeared to engage in activities distinctly different to their daily lives back home. In the case of younger participants, the commonly proposed activities during their visits, namely, drinking, clubbing and shopping, were almost identical to the way they utilised their leisure time back home. Even those on more specific journeys to encounter landscapes, wilderness and tranquillity, and to be on the open road, claimed a strong desire to escape the others and be 'in control' only to retreat back to 'safe havens' (Trauer & Ryan, 2005) during the evening. As Mehmetoglu *et al.* (2001: 20) suggest, travellers are keen to assert their independence by arranging solo trips and attempting to avoid the 'beaten track', but still seek comfortable accommodation and reliable forms of transport. Transportation, or more importantly, personal forms of transportation also appeared to play important roles in the journeys of backpackers to Norway. While those in Oslo and Bergen were mainly happy to go on foot and stay at a singular destination, others found at different locations, suggested that their own vehicles were paramount to their journeys and experiences. Without personal modes of transport, journeys were perceived as being devalued and consequently resulted in a lack of control. For most, the car or motorcycle enabled complete freedom with the surroundings and personal journeys to be constructed without fear of the impact or influence of others. Although many were motivated by the desire to experience Norway's vast wilderness and landscapes, almost everyone conceded that this could not be fully enjoyed and adequately experienced without personal modes of transportation. Landscapes appeared to be a key motivational factor for those with a specific intent to visit Norway and in several scenarios the desire to be alone and to escape was equally as important as the views and vistas themselves.

## Chapter 12

# Town of 1770, Australia – The Creation of a New Backpacker Brand

PETER WELK

## Introduction

In the last 15–20 years, backpacker tourism has emerged from its scruffy 'hippie' and 'muesli' image to constitute an increasingly powerful and profitable economic force on the tourism market. Backpacker-style travel now appeals to wide parts of (still mainly Western) societies, thereby incorporating an increasing share of medium-budget independent travellers. This ongoing trend has major implications for the processes influencing the development of backpacker destinations.

For a long time, the management of businesses in the backpacker industry was dominated by lifestyle entrepreneurs, career changers who had been backpacking themselves in the past and saw an opportunity to combine their passion for the traveller scene with making a living. Since the turn of the century, this pattern has been changing rapidly. With a continuously growing backpacker market and increasing profit margins due to their higher expenditure levels, many managers of backpacker businesses now 'drop down' from the corporate world. They have introduced suits and sports cars, revenues per bed and corporate branding into the previously somewhat anarchistic backpacker business community and have turned a world of mainly sustainability based one-man or family businesses into a veritable industry. Thus, the mainstreaming of backpacker travel and the mainstreaming of back-packer business management are going hand in hand.

We have seen plenty of research being committed to the mainstream-ing of backpacker tourism in recent years. However, the parallel institutionalisation of the backpacker industry has yet to be investigated to fully understand its implications for decision-making processes of young travellers as well as for the tourism development at particular destinations. The extent to which Australian backpacker industry players now exert their market power in order to influence and control these processes will be a central issue in this chapter.

This chapter starts with a discussion of changing travel patterns of backpackers on the Australian East Coast and the impact of a mature and mainstream backpacker industry on the development of backpacker enclaves. This being a case study, the recent developments in an upcoming backpacker hotspot, Town of 1770, and the implications of the artificial hype created around the place by the backpacker marketing industry are investigated. Issues addressed are location, key events that contribute to the destination's increasing popularity, popular activities, demographic changes and the resulting hike in real estate prices. The poor quantity of secondary material available on the destination was enhanced by participant observations during a visit in early 2007, traveller blogs and email interviews with local informants.

## The Australian East Coast Trail

Australia is, probably, the country with the most sophisticated backpacker infrastructure. At more than 3 billion Australian dollars, backpacking generates an unprecedented 20% of the country's international tourism revenues (Tourism Australia, 2008). Apart from the major cities, which (with the exception of Sydney) are mainly visited for 'technical' purposes, such as work, travel logistics and transportation necessities, it is a string of smaller towns along the East Coast between Sydney and Far North Queensland that attracts the bulk of backpackers travelling around Australia. Among these, the primary backpacker destinations are (from south to north) Byron Bay, the Gold Coast, Noosa, Hervey Bay/Fraser Island, Airlie Beach/Whitsunday Islands, Magnetic Island and Cairns. Of secondary importance are Rainbow Beach, Mission Beach and Cape Tribulation (Figure 12.1). Some of these destinations are veritable enclaves, i.e. the backpacker infrastructure is so condensed it forms a bubble within which the backpackers can move without getting into much contact with its environment (cf. Richards & Wilson, 2007) – namely Airlie Beach, the hop-off point for the Whitsunday Islands and their share of the Great Barrier Reef.

Despite the rapid and constant growth of the Australian backpacker scene, the variety of places visited along the East Coast has not increased

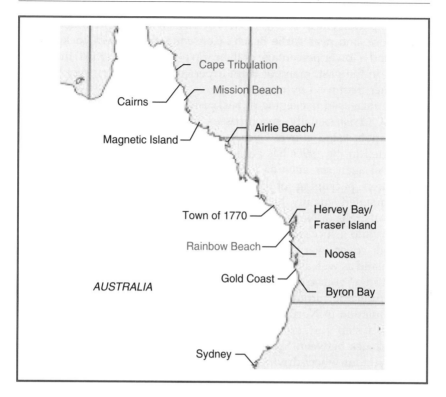

**Figure 12.1** The Australian East Coast Trail

significantly in recent years.[1] Two major factors need to be taken into account in order to understand this:

- The rise of budget airlines, in conjunction with a decrease in average journey duration, has resulted in more selective and predetermined travel patterns. Young people no longer 'drift' aimlessly around Australia, they often determine well in advance the particular places they want to visit, and due to time constraints, they move swiftly between these destinations. The result is an 'overland island hopping' type of travel pattern: backpackers proceed from enclave to enclave without hold-ups in between. This development has been simplified by the advance of low-cost airlines, which enable back-packers with limited timeframes to bridge the vast distances of the continent quickly *yet still* on a budget. Airports frequented by these airlines more often than not can be found in the proximity of

backpacker enclaves (e.g. at Hervey Bay, on the Sunshine Coast near
Noosa, and near Airlie Beach). Consequently, modern backpackers
spend a lower percentage of their travel time 'on the road' than they
did in the past: many of them experience the country spot by spot
rather than road by road.

- The increased packaging of backpacker travel along the Australian
  East Coast usually encompasses only the standard backpacker
  hotspots listed above. Over the last couple of years, a trend that
  started in the 1990s has been intensified by backpacker-specialised
  travel agencies, such as *Peterpans* and *Travel Bugs*: the section-by-
  section packaging of the journey. This means that in heavily
  frequented, mainly highly developed backpacker countries, such
  as Australia or New Zealand, the bulk of travellers now bundle their
  trips into a succession of backpacker-oriented package deals. These
  tours cover every major travelling route in Australia and New
  Zealand as well as much of South Africa. Thus, you can individually
  'package' your trip in several stages all the way from Darwin down
  to Adelaide, around the South and up the East Coast to Cape
  Tribulation in North Queensland. *Peterpans* & Co. have even taken
  this a step further: they 'package' packages, selling a series of
  packages between Sydney and Cairns, which may include a bus
  pass, four-wheel-driving Fraser Island, sailing the Whitsundays,
  rafting the Tully River, doing a bungee jump, and a set number of
  nights in a selection of partner hostels usually operated by big
  chains, as well as free hours of internet in all outlets of the respective
  agency. These all-in-one packages are cheaper than booking the
  single components individually (and they save precious time), and
  the client can still cling to the illusion of travelling independently by
  making his/her own selection among the given hostels – while
  being channelled along a string of industry-controlled hotspots. Of
  course, these packages can be booked online in advance from home,
  ensuring a completely hassle-free journey once in Australia: all-
  inclusive holidays for independent travellers.

These developments have turned the East Coast Trail into the East
Coast Superhighway, both in terms of traveller volume and travel
velocity. Backpackers now rush from enclave to enclave by the
thousands, while the trickle of more independent travellers to rather
off-the-beaten-track destinations has increased at a much slower pace, if
at all. Backpackers' now stronger dependence on commercial, institutionalised means of transport (namely bus companies and budget airlines)

results in a decreasing flexibility in route and destination choice, and many realise only too late the extent to which they have submitted themselves to the constraints of their package deal. Communities not frequented by major bus companies such as *Greyhound* or *Oz Experience* are able to attract only the more explorative backpackers, namely people travelling with their own means of transport, but not the bulk of the market.

## Backpacker Enclaves

The question to be addressed before we get to the case study of this chapter, Town of 1770, is this: how does a peripheral backpacker destination rise to 'hotspot status' in a tourism branch that is as institutionalised, competitive, packaged and, therefore, as inflexible as backpacking along the East Coast has become over the last couple of years? To answer this question, an understanding of the dynamics of backpacker enclave development is necessary.

With few variations, traditional resort lifecycle theories see backpackers as explorers of new, usually remote destinations, retreating from the increased popularity of previous insider tips. They support a basic tourist infrastructure at the new location and a growing market at their place of origin: like in the backpacker novel *The Beach* (Garland, 1996), it is nearly impossible to keep these recent 'discoveries' secret, as word of mouth, online blogs and forums and low-budget guidebooks will spread the word and attract more travellers, which contribute to the establishment of a more sophisticated backpacker infrastructure by local operators, including transportation by boat, bus or private vehicle to and from the enclave. This way, 'these destinations become accessible to the great majority of backpackers for whom the journey primarily represents less exploration than "time out"... Soon the numbers of these visitors increase significantly, with many being attracted by the destination's social reputation at least as much as by other factors such as natural beauty or cultural significance.... Once declared a desirable destination..., visitor numbers can expand almost exponentially' (Westerhausen & Macbeth, 2003: 73). By that time, the 'explorers' will already have left to look for a new, yet more remote destination. Sooner or later, destination awareness and demand in tourist-generating regions will have increased enough for (usually external) tourism players to invest in an upmarket development of the place and to promote it as a more conventional tourist haven, displacing the (local) backpacker facilities.

Mainstream tourism flocks in, backpackers are driven out. Erik Cohen (1982a) did exemplary research on these processes on islands in Thailand.

With the mainstreaming of backpacking, paralleled by the blurring of its boundaries to conventional tourism, these models seem outdated. While the more experimental, explorative backpackers usually still precede a destination's formal tourism development, the bulk of the market does not 'discover' a destination anymore, but is 'pushed' to discover it by (at times) aggressive promotion efforts. For a long time, the backpacker industry was following and feasting on the trends that came up in the scene. Now, it has reached a position where it can determine these trends itself. In Australia, power relations between the backpacker scene and its service providers have thus slowly been turned upside down.

Furthermore, the establishment of a sophisticated backpacker as well as a mainstream tourism infrastructure now often go hand in hand. Therefore, the concept of strict spatial separation between backpackers and conventional tourists (cf. Welk, 2004) is often not accurate anymore. Most independent travellers have started to abandon their 'tourist angst' and mingle with the antitheses of their traveller role models without much ado. Backpackers don't shy away in principle anymore from making use of conventional tourist facilities, from day tours to four-wheel drive rental and private ensuites. On the ground, this has led to a more flexible intermixture of both the backpacker and the more upmarket tourism infrastructures, i.e. backpacker hostels can now be situated in the vicinity of posh holiday resorts without automatically scaring away their clientele. Their patrons might share the dancefloor, a scenic flight or a surf course with mainstream tourists without thinking twice.

## Town of 1770: A Demographic Boomtown

Until recently, the small Town of 1770 and its big sister community, Agnes Water, were sleepy, inconspicuous villages on the Discovery Coast[2] in central Queensland, about mid-way between Bundaberg and Gladstone, surrounded by pristine Bustard Bay and little-known, unspectacular national parks. A backpacker who travelled up the East Coast in 2004 claimed she had never heard of the place. But then, the community has grown at a record pace in recent years. In the year to June 2006, 'Miriam Vale Shire[3] was the fastest-growing Local Government Area in Queensland for the third successive year, with an annual growth rate of 5.7%'. This growth rate meant an acceleration in comparison to the previous five years, which had shown faster growth than the second half of the 1990s. Between 1996 and 2006 alone, the Shire grew by 2100 inhabitants, or 57% (Queensland Government, 2007: 35–36). Another source quotes an 85.5% increase between 2001 and 2006, from 1490 to 2763 people (Osborn, 2006). The new residents seem to be mainly middle-class bohemians who move to the Shire from major cities in search of solitude and tranquillity, many of whom prey on the developing tourist boom – as artists, resort and tour operators.

The population increase has been accompanied by a construction boom. The pristine location has become not only a preferred place to live, but also to maintain a holiday home and to invest in tourism facilities: 'Where wooded hillsides once created a soft green horizon, designer holiday homes owned by absent landlords have appeared. ...Sterile and exclusive apartment resorts look set to dominate' (Footprint, 2004: 301). It is interesting to look at the number of building approvals in Miriam Vale Shire since the turn of the century (Figure 12.2). While the number of approved dwellings remained relatively steady, the 'Other' category, arguably driven by resort development, shows a rapid increase (GAPDL, 2006: 8).

According to the 2005 Gladstone Region Project Status Report, one tourism project had been completed; two were under construction and three under investigation or planned in Agnes Water/Town of 1770 in the year to June 2005 alone.[4] Most of these were resort developments (GAPDL, 2005).

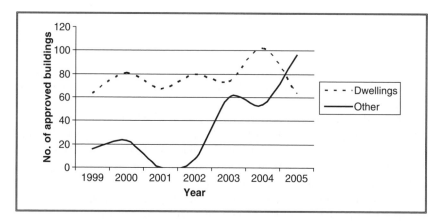

**Figure 12.2** Building approvals in Miriam Vale Shire, 1999–2005 (Own chart, based on GAPDL, 2006: 8)

The result is a gold rush mentality among both external investors and some local landowners. Between 2001 and 2006, median house prices increased by 240% (GAPDL, 2006: 8). In 2004, median real estate prices reached 1.1 million dollars, up from $385,000 in 1999. This development is a source of both wealth and conflict: in early 2007, the owner of a prominent piece of land in Town of 1770 complained that he was unwilling but under strong pressure to sell his 1000-m$^2$ property, which was estimated to be worth two million dollars. 'It seems [that Agnes Water and Town of 1770] have succumbed [to] and are rapidly becoming a prime victim of the great East Coast property phenomenon. As locally predicted, the money has moved in and the locals have moved out' (Footprint, 2004: 301).

In order to avoid becoming a mere 'copy' of other East Coast tourist areas, such as the Byron and Sunshine Coasts, and to preserve the largely untouched Discovery Coastline, the local community 'decided to block what is available to developers' by introducing 'some of Australia's strictest building codes', says Greg, the owner of *Cool Bananas Hostel*. 'Our Council is mainly made up of long-term locals, and they are very conservative. It frustrates the hell out of developers but keeps the pace of change in an orderly and manageable way'.

## Town of 1770: A Rising Star in Australia's Backpacker Universe

Despite the boom, the Discovery Coast is still a peripheral tourist destination off the beaten track and major transport routes as tourism started here only recently with a 'blank canvas' (*Cool Bananas'* Greg). Until recently, hardly anyone knew the place existed, even among backpackers. With their small population and at a distance of about 60 km from Bruce Highway, the two settlements can be found only on more detailed roadmaps. The distance to both Bundaberg and Gladstone is approximately 120 km.

Currently, the destination receives about 25,000 tourists per year – 'mainly flush young backpackers and gray nomads with a good dye job' (Ford, 2007) – and a fraction of the 650,000 that visit Airlie Beach (Worldtourism, 2008), which equals Agnes Water in population size. Although Agnes Water is about ten times as big as Town of 1770 with nearly all tourist facilities located there, the destination is being marketed by the catchier brand name 'Town of 1770' – or simply '1770'. 'Since there's not much there [in Town of 1770], most people stop in Agnes Water and pretend', comments an anonymous blogger.

During a stay in Brisbane in mid-2006, the author frequently heard backpackers talk about Town of 1770, while stickers and flyers were omnipresent in hostels and backpacker travel agencies. It seemed to be 'one of the most talked about locations on the east coast' (Dunn, 2004), and everyone had the feeling they had to go there 'before it's too late', i.e. before it started to look as bland as other East Coast backpacker havens, such as Airlie Beach or Hervey Bay. 'It was nice to enjoy the location while it is still young, as *Let's Go* Australia points out that *it's going to be huge'* (Dunn, 2004). The 2007 Australia issue of *Lonely Planet*, compiled in the same year, has a quarter page full of similar references on the destination: 'surf away from the crowds... at a healthy distance from the beaten track... a refreshing lack of crowds... get away from it all' (Lonely Planet, 2007: 369).

Being curious about this new spot on the backpacker map, I travelled to Town of 1770 personally in January 2007. What struck me most, apart from its ideal setting as a backpacker enclave, was that, despite it being the high season, barely any backpackers were actually staying there. I had the impression that all the hype and promotion that engulfed the placename elsewhere was merely hot air – much ado about nothing. In terms of a backpacker-friendly infrastructure, there were only two campgrounds and two hostels: *Cool Bananas* and the *Southern Cross Tourist Retreat*, which had only opened in December 2006.[5] Not even the backpacker bus company *Oz Experience* stopped there, although rumours in Brisbane had said so.

From Brisbane, 1770 had not looked like an insider tip; after all, its promotion was plastered all over hostels and backpacker tour agencies. I realised that in this case, the usual pattern was turned upside down: elsewhere, you may believe that a destination is an insider tip, until you get there and realise it's as crowded as any other hotspot. But then, 1770 did *not* feel like one – until I got there. I couldn't help the feeling that the tail was wagging the dog at the place. Thus, in terms of quantity of travellers and backpacker businesses, 1770 *was* off the beaten track. But in terms of its quality, i.e. its awareness in backpacker minds and presence in marketing, it was *on* the beaten track. So this discrepancy between 1770's perceived popularity and actual frequentation made me even more curious, raising a range of questions: why does the backpacker industry promote the destination so fervently despite its lesser importance and lack of facilities to accommodate large backpacker quantities? Why is this happening now and not much earlier? Is the situation in 1770 comparable to other backpacker destinations, and to what extent does the community have the power to avoid the mistakes made elsewhere in the past? The answers to these questions are detailed below.

## Location

A look at the Australian East Coast map reveals an uneven distribution of the prime backpacker hotspots (see Figure 12.1). Most striking is the 900-km gap between Hervey Bay/Fraser Island and Airlie Beach/Whitsunday Islands. For backpacker transport operators and their clients alike, this poses a dilemma: a 15-hour bus journey and the gnarling feeling of missing out on something. While 1770 is not exactly located mid-way in this gap, it nevertheless cuts this distance short by 200 km. As my informant, Tracey at *Peterpans*, states: 'We didn't want our customers to rush between these places. Stopping at 1770 allows them time to "chill out" between Fraser and the Whitsundays'. Its fair distance of 60 km from Bruce Highway should not be considered a significant disadvantage if we take into account that other well-established backpacker destinations, such as Airlie Beach (23 km), Hervey Bay (39 km) and Rainbow Beach (98 km) aren't exactly close to major highways either.

## Activities

Despite its laid-back character, 1770 offers a wide range of activities to visitors, catering to the interests of most independent travellers:

*Surfing*: Agnes Water is recognised as the northernmost spot on the East Coast suitable for surfing. The waves are not of the kind that gets dedicated surfers' hearts pounding faster, but is all the more suitable to take surf lessons.

*Diving*: On the Discovery Coast, Australia's northernmost surf beach meets the southernmost point of the Great Barrier Reef, thus it is the only place in Australia suitable for surfing *and* tropical diving. 1770 is the closest mainland settlement to Lady Musgrave and Lady Elliot Islands, which are rated among the best diving spots along the Reef. A daily boat shuttles ship diving and snorkelling enthusiasts to Lady Musgrave Island and back.

*Riding Scooter Roos*[6]: One of the first tour operations aimed at young travellers in 1770 is still among its most original and popular: fake, i.e. moped choppers can be hired individually or by joining a daily tour around the area. The popularity of the *Scooter Roos* has most recently sparked *Street Beat Cycles*, which add aquachoppers, quadracycles and chopper cycles to the range of local transport choices.

*Riding the LARC*: The *Lighter Amphibious Resupply Cargo* vessels, a kind of amphibious bus used by the US Army in the 1960s, now serve as unconventional tour vehicles across the sand bars and inlet at Bustard Bay.

*Relaxing on the beach*: Agnes Water's main beach is long enough for the few visitors to disperse and occupy their own private niches. Currents are fairly safe for swimming, and the beach is within walking distance from anywhere in town.

*Watching sunrise and sunset over the sea*: Due to its location on a headland jutting out into Bustard Bay, Town of 1770 is one of the few places in Australia to view the sun both rise and set over water.

*Scenic flights*: The full beauty of the area only unfolds from above. Judging by two traveller blog entries, the scenic flights offered from Agnes Water airstrip are a hit not only among more affluent tourists, but also among backpackers, despite being costly for someone travelling on a low budget.

*Visiting nearby national parks*: Three national parks surround Town of 1770: Deepwater National Park along the coast to the south, Eurimbula National Park along Bustard Bay to the north and Joseph Banks Conservation Park, which encompasses the headland around Town of 1770 proper. Accessible only by one's own means of transport, and at times only by four-wheel drive, depending on road conditions, the two national parks still receive very few visitors, but serve as a buffer zone against further development of the surrounding area: the Discovery Coast features the longest stretch of primordial coastal wilderness between Sydney and Cairns. According to *Cool Bananas'* Greg, only eight out of 250 km of coastline are open to construction development.

### Key events

Twenty years ago, Town of 1770 was 'hardly developed', accessible on 'a rough dirt road', with only one motel, campground and caravan park (Wheeler, 1986: 256). However, Tony Wheeler, founder of *Lonely Planet* and a thorough observer of backpacker dynamics, already realised the potential of the place at this early stage: 'It appears that this area may become a lot more developed in the future so see it while you can' (Wheeler, 1986: 256). Until 1999, Agnes Water did not feature more than a petrol station, a few shops and still one tourist resort (Rough Guide, 1999: 450). So when did these two sleeping beauties start to wake up to become a mentionable name on the East Coast tourist map? No local informant could pinpoint a certain time or event that sparked the place's rise as a mentionable tourist destination. It seems that it was rather a range of developments that made the wider tourism industry public aware of the place's potential and attraction:

*1970: Round Hill is renamed Town of 1770.*[7] In commemoration of James Cook's first landing on the Queensland coast, the residents of what he once called Round Hill renamed the spot Town of 1770. Many years later, when tourism development started trickling in, the catchy name proved to be an ideal token for the branding of the whole destination.

*1993: 1770 Festival established.* 'From humble beginnings, the *1770 Festival* has grown into a significant regional event attracting thousands of visitors from intrastate and interstate' (Let's Connect, 2007). What started as a re-enactment of the landing of the *Bark Endeavour* crew has turned into a four-day festival including an art exhibition, a comedy show, a street parade, the *Southern Cross Film Festival* and, from 2009 on, a Jazz festival (1770 Festival, 2008). In the meantime, 1770 also hosted other supra-regional events, such as the *AussieHost* Customer Service Programme.

*Late 1990s: Backpacker travel agencies start including 1770 in their package deals.* According to my *Peterpans* informant, this started in 1998. However, it is questionable that this destination was on their schedule at a time of unsealed roads, a lack of public transport and the absence of backpacker accommodation (see below). Yet, the inclusion of 1770 in East Coast packages was a major step to becoming a backpacker centre. In the first place, it raised destination awareness through its presence in brochures and promotion. But it also lifted the place up to eye-level with other backpacker hotspots.

*2000: The road from Miriam Vale gets sealed.*[8] While the previously unsealed road was in a good, all-weather condition for all types of vehicles, psychological compunctions have to be considered in the estimation of this measure: on the map, an unsealed road suggests difficult accessibility and little tourist appeal. (While adventurous, anti-tourist attitudes are still widespread in the traveller scene, only a minority of backpackers will put them into action by actually venturing off the beaten track.) More importantly, overland bus companies hesitate to send their coaches on gravel roads, and most car rental agencies don't allow their customers to touch dirt roads, either. Thus, the sealing of a road is an important contribution to opening up a destination to both backpackers and conventional tourists.

*2003: Cool Bananas Hostel opens.* Without low-budget accommodation, a destination is lost to a backpacker scene. Campgrounds are only accessible to people travelling with their own (or rented) vehicle – a minority among backpackers – so the establishment of a hostel is a milestone to becoming a mentionable backpacker destination. *Cool Bananas* is frequently mentioned as one of the country's best in traveller blogs and can therefore be considered an attraction in itself.

*2004: Scooter Roos starts operating.* It seems that the founder of the *Scooter Roo* chopper and bike tours was one of the driving forces of backpacker-oriented experience tours in 1770. Until the present day, *Scooter Roos* is one of the unique selling propositions in town, i.e. an attraction not to be found elsewhere.

*2005: Greyhound starts servicing Agnes Water.* Had a local bus from Bundaberg serviced Agnes Water in the past, the real breakthrough came with its twice-daily inclusion on the timetable of overland bus company *Greyhound* in 2005. Later, backpacker bus company *Oz Experience* wanted to jump on the bandwagon by trying to strike a cooperation deal with *Cool Bananas*, but was turned away. *Oz Experience* now stops at Miriam Vale, from where backpackers can order a pick-up service.

*Mid-2000s: Blogging becomes a mass phenomenon.* With the mainstreaming of the Web 2.0 and the advance of social networking sites, backpackers themselves started promoting Town of 1770 by sharing their experiences with the wider online community. A look at *Google* search results reveals the impact blogs in particular have on sheer destination awareness: in January 2007, the search engine returned about 3000 results for 'Town of 1770'. By January 2008, this figure had increased to 32,600; in February 2009, it stood at 46,000.[9] At the time of writing, the social networking site *Facebook* featured 17 groups dedicated to the destination.

*2009/10 (ever?): The Discovery Coast gets its own airport.* Currently, Agnes Water has only a small airstrip for scenic flights and local shuttles to surrounding airports. Back in 2006, there were plans by a Sydney-based investor to construct a 'commercial airport, within 15 minutes' drive from Agnes Water, to be built and operating by 2008' (Singer, 2006). According to local informants, these plans have been dropped in the meantime due to local residents' resistance and the fact that there are airports available in nearby Bundaberg and Gladstone.

As yet, backpacker facilities are not part of major tourism infrastructure projects in Agnes Water and therefore cannot be blamed for the real estate hype. However, the big rush hasn't kicked in yet, and it would be naïve to uphold the outdated image of backpacking as low-scale and low-impact tourism. The backpacker industry is not only observing, but actively trying to push forward the development in 1770, and once they get a viable mainstream market to head for the destination, major backpacker hostel corporations, such as *Nomads* or *Gilligans Island*, might be waiting in the wings to invest in multimillion-dollar, purpose-built backpacker resorts at prime beachside locations.

*Cool Bananas* owner Greg provides an insight into these thrusts:

> I have refused to be part of organised package deals even though we are offered them all the time by the large travel agencies (and their promises to make us "rich").... *Oz Experience* wanted to use *Cool Bananas* as their own exclusive hostel. The contract was worth several hundred thousand dollars, but I refused. ...So they had the shits that we dared to refuse an industry heavyweight like them and told us that they would wipe this town off the map, or something along those lines.

The compromise was for the buses to stop at Miriam Vale, from where backpackers have to order a connection shuttle, run by a local entrepreneur. However, on its website, *Oz Experience* now recommends *Cool Bananas* as *the* place to stay in 1770 (Oz Experience, 2008). Whether this is a new strategy to endear themselves with Greg or a means of retaliation by 'infiltrating' his hostel with party backpackers and 'their wheelie suitcases and parent's credit card' (Greg) is up for speculation.[10]

## The Byron Bay Syndrome

Apart from Airlie Beach, the northern New South Wales backpacker hotspot, Byron Bay, is a frequent destination of reference in forecasts on what Town of 1770 could be heading for. Once a centre of Australia's 'hippie' movement, the town is now a posh Bohemian haven with its spectacular setting on the easternmost point of the continent, its surf beaches, the lush, green hinterland and a liberal social environment. Needless to say, the place is one of the most popular backpacker destinations outside of the big cities: in 2007, Byron Bay (as well as Airlie Beach) was visited by 142,000 international backpackers, one third of all backpackers to regional Australia, who spent one million nights there (Tourism Australia, 2008). Whether or not Byron Bay has succeeded in going its own way in tourism development is an issue of fervent debate. Seeing what was happening less than 100 km to the north on the Gold Coast, the community decided in the early 1990s to cap both unrestricted population growth and tourism development in order to preserve the unique authentic ambience of the place and not to exceed the carrying capacity of its environment. Greg, who owned a hostel in Byron Bay before he 'escaped' to 1770, argues that: 'Byron Bay has been "loved to death" – 1.2 million tourists a year in a town with a permanent population of 10,000. So-called "backpackers" are a major part of this deterioration supported by greedy operators who have no regard for the aesthetics, ethics, or the longevity of what sustains them. . . . Five hostels are owned or part-owned by one single family'. Greg, who says he found in 1770 what went amiss there, claims that here

> most locals appreciate that an economy has emerged around the backpacker business and find them pleasant in general. Organised tourism is only about five years old here so people aren't over it yet. Another very important aspect. . . is to promote businesses that are owned by individuals, not by some large operator like elsewhere. . . . This shares the rewards of tourism around the community. . . , and that happens here very much – for now.

The question to be asked here is: can backpackers, who are usually seen as mainstream package tourism's 'Trojan Horse', serve as a stronghold against it? In their Byron Bay study, Macbeth and Westerhausen (2003: 83) claimed that 'backpackers and alternative communities in particular appear to be natural allies in the struggle for sustainability and local control. Backpackers provide a ready market in which authenticity, individuality, colour and nature are appreciated.' But they also acknowledged that 'tourism enclaves frequently develop as a consequence of market forces instead of formal planning decisions' and that 'the popular press and publications like *Lonely Planet* have a stronger influence on visitor numbers than Shire Councils' (Macbeth & Westerhausen, 2003: 82).

In the years that have passed since their Byron Bay investigation, the Australian backpacker scene has been drifting further towards mainstream tourism, while the backpacker industry has rapidly professionalised itself. It is now at eye level with other powerful players in tourism, using the same sophisticated strategies to pursue its more commercial interests. The consequence for local communities is that the Byron Bay strategy of promoting itself as a backpacker destination in order to keep mass tourism at bay cannot be applied anymore, because backpacking has become a variety of mass tourism itself.

## Conclusion

Recent years have seen a development that is somewhat new in backpacker tourism: Town of 1770 is being heavily promoted as a new, independent traveller destination, a mixture between the hotspots of Airlie Beach and Byron Bay. In this way, the Australian backpacker industry is turning upside down an old principle of the backpacker resort cycle: in the case of 1770, its marketing machinery precedes the usual process by trying to take over control from the explorative travellers and local entrepreneurs. This is what sets Town of 1770 apart from conventional resort cycles: the marketing industry is promoting a blueprint; it tries to skip the lengthy insider tip stage by putting the cart before the horse. This means a major shift in who's driving the development of backpacker destinations.

Backpackers are fervently picking up on the hype while the place still enjoys off-the-beaten-track status: destination awareness is very strong within the scene, and everyone is under the impression that they have to go there 'before it is spoilt by the crowds', which results in accelerating this very process. Effectively, bus companies *Oz Experience* and *Greyhound* are both servicing 1770. Real estate prices are soaring and already overheating, and the backpacker industry is abuzz with anticipation of the demand it is creating here.

However, the reality in 1770 itself does not yet reflect this hype. Despite the parallel emergence of upmarket resort tourism, the pace of development is much more modest here than elsewhere along the coast of Queensland, due to strict control by the local Council. Having learnt from the experiences made in Byron Bay and on the Sunshine Coast, the residents of Town of 1770 are resisting the influx of major players of the backpacker and tourism industry.

Obviously, the backpackers themselves don't play a consciously active role in this power struggle. They are transient onlookers without the necessary insight to fully understand the scope of developments unfolding in 1770. They may have a premonition that the good times may be over soon, but they don't have to bear the consequences of their visit, each one adding a footprint to the widening trail to Bustard Bay. This way, they unwillingly (and often enough indifferently) become agents of the wider backpacker industry.

Thus, despite its inconspicuous remoteness, Town of 1770 gives the impression of a community under corporate siege. While the community pursues low-scale, sustainable backpacker and resort development, it is under extreme pressure by the tourism and backpacker industry to open up to large-scale investment: internet cafés, pubs, more surf schools, more dive operators, nightclubs, pub crawls, a bus station serviced four times a day, the branch office of a backpacker travel agency, a constant stream of free-spending visitors – the lot. This pressure is rarely executed directly by industry representatives such as in the example given above, but more subtly through the inexorably increasing demand they are fuelling among their clientele – and the myth of 'you got to see it before it's too late'. The destination is still in the early stages of becoming a backpacker hotspot, and may remain so for a couple of years more. However, with aggressive, self-conscious and professional backpacker industry marketing being a new, powerful factor in spreading the word and increasing the hype, the development from pristine, remote community to fully-fledged party enclave may proceed much faster than was the case in other destinations, not leaving much time for local residents to keep enforcing their interests.

We have to bear in mind that most backpackers, on arriving in Australia, do not know much about the character and popularity of particular destinations outside of the major cities. This means that the strong presence of a destination in marketing can easily lure them into believing that a minor 'star' such as 1770 is a must-do place to go – as pointed out by *Peterpans* interviewee Tracey – and that heavy promotion can 'blow up' its perceived attraction. In the end, a review of traveller blogs shows that the vast majority of backpackers who actually end up travelling to 1770 enjoy the tranquillity and un-agitated solitude of the place. Thus, they visit the destination expecting a backpacker haven and then enjoy it because it isn't – which is one of the insoluble contradictions of the backpacker scene.

## Notes

1. An exception is Harvest Trail destinations, such as Bundaberg or Bowen, which do not gain their appeal through a location near natural attractions or the presence of a vibrant backpacker scene, but have become important destinations for working holiday makers because of the work opportunities in the local agriculture.
2. 'Discovery Coast' is an artificial marketing term exclusively used for Town of 1770 and Agnes Water, including the adjoining national parks. In fact, it is part of the longer Capricorn Coast stretch.
3. About 65% of the Miriam Vale Shire population are residents of Agnes Water and Town of 1770.
4. Curiously, the 2004 zoning plan for Town of 1770 did not appoint any tourist commercial zones on the headland (State of Queensland, 2004).
5. By late 2008, there were three hostels, including *1770 Backpackers*, a flashpacker-style resort.
6. Various spellings apply even on their own website.
7. Most sources are consistent on this date, but one source puts the renaming as early as 1936 (Discovery Coast Tourism & Commerce, 2006).
8. One source claims this didn't happen until late 2004 (Footprint, 2004: 312), but the road's condition in early 2007 suggested that it must have been sealed earlier than that.

9. It has to be pointed out, though, that this number can vary seasonally between different country sites of Google, i.e. google.com.au, google.de, etc.

10. Since early 2009, however, *Oz Experience does* include an overnight stop at *Cool Bananas* in its programme (Oz Experience, 2009). Thus, the bus company finally seems to have made Greg an offer he could not resist.

## Chapter 13

# A Clash of Cultures or Definitions? Complexity and Backpacker Tourism in Residential Communities

ROBYN BUSHELL and KAY ANDERSON

## Introduction

Students and youth have long been a significant segment of travellers, representing one in five international travellers (Richards & Wilson, 2004) across the global tourism industry. Many of these young travellers are identified as backpackers and can be highly sought after by destinations because of the economic yield associated with this market segment. Backpackers tend to stay much longer in Australia (for an average of 71 nights compared to 25 nights for non-backpacker visitors) and spend on average 50% more per person than other visitors (TRA, 2008). While defined by government agencies as 'a person who spends at least one night in either backpacker or hostel accommodation' (TRA, 2008), these visitors, most often, do not spend all their stay in backpacker accommodation. That travellers classified as 'backpackers' seek alternative and multiple modes of accommodation complicates accepted notions of backpacker identity and behaviour. The age profile of the backpacker, too, is variable and in Australia this profile has been shifting for the past few years. The backpacker market has been predominantly 20–29 year olds (43% in 2000). However, visitors aged 40–49 years now represent 12% (62,000 visitors) of international 'backpacker' visitors to Australia, up from 8% (38,000) in 2000 (TRA, 2008). This suggests that discourses surrounding the term backpacker should not assume synonymy with youth. Stereotypes, such as Cohen's 'drifter tourist' (1973) – a young person with a backpack staying in cheap lodging – have for some time been insufficient to encompass the broad spectrum of backpacker identities, behaviours and purposes. Indeed, contemporary independent travellers and their travel cultures are giving rise to new traveller

187

identities and infrastructures, taking in a diversity of categorisations from the so-called 'grey nomad' to the 'gap-packer' and the 'flashpacker' (Maxwell, 2008).

In this chapter, we take the *im*precision surrounding the term 'back-packer' and its recent offshoot 'flashpacker', to highlight the increasingly complex definitional and management challenges of the independent traveller market in Sydney, Australia. Drawing on findings from a 2005–2008 study of six local government areas that are popular backpacker destinations, we suggest the Sydney case reflects an increasingly affluent traveller cohort with associated pressures for more diverse accommodation options. More likely to carry a laptop and a trolley-style suitcase than a tent and a backpack, this sector pursues leisure and work in a mix that has been increasingly enabled through changes to Australian immigration regulations. The growth of this sector has not, however, seen any diminution in the flow of a set of international travellers to Sydney thought by many residents to conform to much more conventional stereotypes of the young, male, European, hedonistic backpacker. Attracted to the sun and surf of 'sentient Sydney', these still predominantly UK/Irish (21%) and other European travellers generate much heat in the city's coastal residential communities. But, just as the traveller market is itself increasingly differentiated, so too are the responses of the various stakeholders in Sydney to the 'backpacker phenomenon' both diverse and complex.

Modalities of accommodation and age demographic are not the only aspects of definitional ambiguity and blurring concerning the term 'backpacker'. The contemporary backpacker crosses and confounds many conceptual boundaries, including the notion of 'holidaying'. The Australian Government actively encourages international visitors to stay longer via a Working Holiday Maker (WHM) visa. The scheme, 'allows working holiday makers to have an extended holiday in Australia by supplementing their travel funds through incidental employment and to experience closer contact with a local community' (DIMA, 2008). Regarded a strategic success by the Australian Government, the scheme has recently been extended to expand the list of eligible countries of origin, to allow for a stay of up to two years, and to enable visa holders to remain in the same job for up to six months. It is both a labour market initiative and tourism marketing scheme, seeking to enhance economic benefit through long-stay visitors and contributing over AUS$1.3 billion per annum (based on figures of 80,000 WHM visas in 1999–2000, Harding & Webster, 2002). In 2007–2008, 154,148 visas were granted, with 33.3% to UK and Irish residents (DIMA, 2008). Visa holders undertake seasonal agricultural work in regional Australia and a variety of service industry and health sector jobs in major capitals. The contribution of backpackers to local industries is such that the City of Sydney, for example – as one major council area in a city that attracts billions of dollars annually from many forms of tourism – depends not only on these people's patronage of services and facilities, but also their labour (Backpacker Operator, City of Sydney and City of Sydney Council staff, in Allon *et al.*, 2008).

In spite of such benefits to local and national economies, and regardless of whether the travellers are labelled as 'back' or 'flash' packers, their intrusion into everyday life, day and night, is not viewed as all that 'flash' by some Sydney residents. The 'backpacker phenomenon' has become a complex mix of holidaying, travelling, working and residing (Allon *et al.*, 2008), with the backpacker often the objectified visitor or 'other' to whom much is attributed. The term 'backpacker' is frequently used as a pejorative label by residents, many of whom, as shall be discussed later in the chapter, hold such visitors in considerable disdain.

## Challenging our Understanding of Who 'Belongs' Where

In the Australian context of more liberal regulations governing 'working holiday makers', the notion of 'visitor' expands to include someone with a temporary work visa who may reside in one place for many months, and even up to two years. Not only does this complicate definitional boundaries, this style of 'tourist' experience becomes entangled with that of a residential experience, as these (mostly) young people seek to 'feel at home' in rental rather than backpacker/budget accommodation in Sydney (Allon *et al.*, 2008). In cities that aspire to be global, like Sydney, many suburbs are not only desirable to residents, but attractive to domestic and international visitors by virtue of their proximity to excellent surfing beaches, engaging natural and cultural environments, a reasonable stock of rental properties, extensive choices in shopping, dining and diversity of leisure activities and entertainment. As mobilities of all sorts become increasingly commonplace, people's understanding and acceptance of who 'belongs' and whose rights deserve respect, is challenged. Urry (2006: vii) talks, in this respect, of 'contingent mobilities'. He states: '[p]laces are intertwined with people through various systems that generate and reproduce performances in and of that place. These systems comprise networks of hosts, guests, buildings, objects and machines that contingently realize particular

performances of specific places'. The *place* of the backpacker within the
constitutive space of local residential communities is both contested and
little understood.

To better conceptualise the impacts of backpacker tourism in residential communities, ethnographic fieldwork in six local government areas
of Sydney that are popular backpacker destinations explored the
different identities and transformations of place at play in these suburbs.
The study, supported by the Australian Research Council and conducted
during 2005–2008, drew on in-depth focus groups, interviews, discussions and two online surveys, one specific to residents, the other to
backpackers. Many stakeholders contributed, including local residents,
local businesses including backpacker operators, members of backpacker
industry bodies, council compliance and planning staff, elected council-
lors and backpackers within each of the local government areas – the
City of Sydney, Randwick, Manly, North Sydney, Waverley and Wool-
lahra. The Councils were partners in the investigation that also involved
several state government agencies concerned with tourism, planning
regulations and health and safety issues. The research sought to increase
our understanding of the social and cultural impacts of the backpacker
phenomenon.

The work demonstrates that simplistic categorisations based on
arbitrary designations such as 'tourist' versus 'local', or outmoded
concepts of 'communities' as fixed and bounded entities, deny the
mobilities, dynamism and transformations of place that are constantly in
performance and production within a globalising city such as Sydney.
The very language of 'local resident' and 'tourist' sets up binaries and
divisions that bring with them expectations and prejudices. The backpacker and especially the working holidaymaker symbolise the tran-
siency and complexity of mobility, and underscore the socio-spatial
relationality between the global and the local. The backpacker thus
highlights tensions between the fixity and fluidity of communities
through which bodies move on a daily basis as part of the production
of everyday life (Allon *et al.*, 2008a, 2008b). At stake here is not just the
consumption of place by the tourist as they transit through it, but of
people moving constantly for work, leisure, education and social
mobility. The often difficult relationship between the long-stay 'backpacker' and all that is considered 'local' signifies the contested rights and
conflicting purposes of an increasingly diverse and mobile citizenry. The
backpacker embodies a 'clash of cultures' no less – as different lifestyles
and values intersect and mingle; the transient populations and the more
permanent; the unencumbered youth versus mortgage-bearing white

collar families past the fast-living phase of their lives and more concerned with property values; a fast-growing demographic of urban retirees seeking a peaceful and relaxing, rather than exuberant, ambience.

Long-stay transients are themselves creating new forms of permanent space akin to 'ghettoes' that are resented by those who claim ownership of place identity and character. These enclaves (including pockets of Bondi, Coogee, Pyrmont and Kings Cross in Sydney) serve as functional resting and meeting places for independent travellers, and thus become destinations in themselves. Indeed, it is clear from the backpacker responses to the Sydney research that many of these young people travel for the experiences of the enclaves and for social interaction with other backpackers. Wilson and Little (2008) describe this as the 'suspended experience' of traveller enclaves, where backpackers carve out a space that is neither 'here' nor 'there' and, as such, offers *both* the titillation of the 'foreign' and the safety of the 'known'. This desire to mix with other travellers was borne out in the focus groups for the Sydney research. 'If it just was a community that didn't have amenities for backpackers, you'd probably stay away. Like, there are loads of suburbs in Australia that we'll never go and visit because it's suburban' (Bondi Backpackers Focus Group, 2006).

Backpackers are often associated with binge drinking, noise, lethargy and even apathy. Such sentiment was expressed in resident interviews and focus groups by people who felt their right of management over place image and character was under siege: '...To me, "backpacker" just means purposelessness. I mean look at them. They just come out here and they're not actually doing anything. They buy a slab of VB [Australian beer], take it back to the hostel and get absolutely pissed. They're not doing anything' (Bondi resident, Mixed Council Focus Group, 2006); and from another resident, 'What we experience is, for want of a better word "grog tourism"' (Coogee resident, Mixed Council Focus Group, 2006). In some localities, backpackers have come to be so consistently associated with particular behaviours that they have become scapegoats, blamed for the anti-social behaviour associated with ex-cessive alcohol consumption by other social groups:

> Sometimes we have to investigate what is said to be a backpacker complaint, and it's found to just be a group of locals, and it might be students or it might just be young people sharing a house, and they're not in the *true* sense "backpackers". It's just that their drinking and their partying is the same type of behaviour that people attribute to backpackers, but it's not truthfully a backpacker

issue. So, the term "backpacker" might be *unfairly* applied sometimes, but in the majority of residents' perception it's basically a big
problem in Bondi. (Compliance staff, Waverley Council Focus Group,
2006)

Abandoned cars in Bondi were also regularly deemed the result of
thoughtless backpackers, though one study found that around 80% were
in fact Australian-based owners (Waverley Council, 2002). Council
officers and backpacker operators acknowledge that local residents and
domestic visitors are often responsible for problems that are attributed
to backpackers. Focus groups conducted for the Backpackers in Global
Sydney study noted that many backpackers, even more affluent ones,
are simply too frugal to dump cars (Backpacker Operators Focus Group,
2006).

That travellers are frequently confused with international and local
university students and/or local youth generally, that the behaviours of
'locals' can be misrecognised and misattributed as those of 'backpackers',
and that long-stay transients are conflated with 'backpackers', regardless
of their semi-permanent accommodation or employment status, belies
the difficulty of furnishing concrete and accurate definitions of backpackers in context.

So, what criteria do either the casual observer/resident or the official
statistician and regulator use to determine the category to which such
people belong? For official purposes, the international visitor is classified
by the type of entry visa issued – tourist, student, temporary resident,
working holiday maker. For statistical purposes, data is collected from
commercial accommodation providers and 'backpackers' determined by
Tourism Research Australia as 'spending at least one night in either
backpacker or hostel accommodation' (TRA, 2008). For the casual
observer, it is likely to be a combination of age, ethnicity, accent,
appearance and behaviour that informs opinion, as though there were
some 'typical backpacker', just as there is (an imagined) 'typical local'. To
residents, '[t]hey just party on and off all day, night, whatever. They have
no furniture, no belongings except a backpack with a few things in it.
They all have an accent of some description, and they have no regard for
the building at all. They come and go. They're here for a good time...'
(Pyrmont Resident Interview, 2006). Many Council staff interviewed
during the research – social and community planners, building regulations and compliance officers, planners, research staff – expressed views
that point to this ambiguity: '[S]ure, they're backpackers... but they're
here for three months... they're like residents'; 'are they a backpacker or

not? I suppose they are really, but they're not... They're like everyone else renting a place'.

The backpackers in the Global Sydney research project sought to bring into focus this fuzziness of perception and definition around accommodation, types of people and belonging – definitions that not only inform community attitudes, but also government policy and regulatory frameworks. The research makes explicit the considerable perceptual differences among stakeholders – residents, local business, industry groups, local councils, state bodies and visitors. It has brought to the fore the fluidity of categories and places and highlighted the sense in which they are all in process. It also demonstrates the normality of movement *through places* – and the impact of mobility on various expectations of quality of life within resident communities. The backpacker is part of the narrative of the construction of place where many paths, purposes and interests intersect as (dis)junctions between the local, national and global. The challenge for local governance bodies is a seemingly impossible juggling act, as these different forces and stakeholder needs all seek consideration in the management of places for economic growth, local prosperity and harmonious community relations. Government at all scales encourage local business enterprise and visitors.

The WHM visa makes it easier to stay both legally and financially, and the infamous Tourism Australia campaign[1] of 2006 'So Where the Bloody Hell Are You?' was a clarion call to the young traveller. International backpacker numbers have been growing faster than other international visitor segments to Australia and were targeted as a key niche market for development in the Tourism White Paper (Commonwealth of Australia, 2003). As many as 66% of the 309 surveyed backpackers in the Sydney research were travelling on a working holiday visa. The provision for lengthier employment here resulted in an overall increased length of stay in Australia, comprising interspersed periods of holiday and work. It also increased backpacker expenditure. Of the $2.8 billion in backpacker tourism revenue in 2002 in Australia, $2.5 billion or 88% came from international visitors (Heaney, 2003) – on average spending more than double that of non-backpackers (TRA, 2008). The majority of the international backpackers to Australia are from the UK, Germany and 'Other Europe', accounting for 63% of the visitors (and 64% of international backpacker visitor expenditure) in 2003 (Ipalawatte, 2004) and 53.8% of international backpackers in 2007. Although a much smaller share of the market, backpackers from Asia are growing in number, with Korea and New Zealand having the strongest growth, both increasing 11% per annum since 2000 (TRA, 2008). The majority (84%) of international backpackers are 15–34 years of age. The gender split is equal (TRA, 2005), although in the Sydney research, male travellers made up 60% of the Backpacker Survey responses – a proportion at odds with the growing international trend for female travellers to outnumber males in the youth travel market (see Richards, 2007: 5; Wilson & Little, 2008). Seventy-two percent of international backpackers to Australia travel alone (Ipalawatte, 2004). The backpacker segment of the tourist market to Australia is also more resilient than most, noted for maintaining solid growth during periods of international uncertainty (Heaney, 2003). Many Europeans, well-travelled in their continent, choose Australasia as their first long-haul destination before tackling less-developed destinations (Richards & Wilson, 2004), but 2008 marked a decline in first-time visitors (down to 51%) for the first time in seven years (Munro, 2008). Despite this, the trend for overall growth continues with 3% growth rate per annum since 2000 (TRA 2008). The domestic backpacker market has also grown since 2000, especially for persons aged 40–49 years (TRA, 2008).

## Defining Difference… Slippery Subjects

All definitions of travel presuppose concepts of space and time – movement away from home and routines for some time, just as time is an important measure by which someone comes to be regarded as local (Wilde, 2006). Another variable that impacts significantly on the definition and perception of the 'backpacker' is purpose: holidaying as opposed to working, travelling as opposed to dwelling. Tourism is primarily a leisure-oriented activity, but the boundary with other activities is highly permeable and hence there is definitional slippage in this regard as well. Hall (2005b, cited in Wilde, 2006) presents a model (Figure 13.1) of temporary mobility, integrating tourism with other forms of mobility from the briefest hours shopping, to many months of extended working holidays, to temporary migration (Cole *et al.*, 2005).

The access to employment enhances the traveller's budget for more time 'away', and affords funding to experience types of accommodation beyond the conventional hostel – the more expensive or those that have lengthier minimum stay requirements. Rented accommodation accounted for 28% of nights spent in Australia by international backpackers in 2007 (TRA, 2008). For most backpackers, renting units corresponds with the period(s) when they are 'living a normal life and temporarily resident, so, getting up and going to work just like everyone else' (Bondi backpacker Interview, 2006). Several of the backpackers interviewed distinguished between periods when they were travelling,

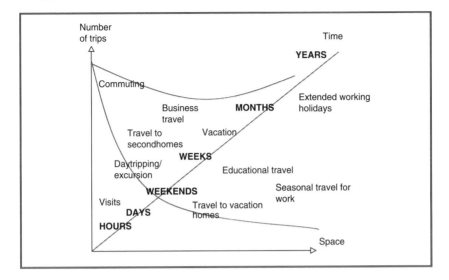

**Figure 13.1** Time: Space compression
*Source:* Hall (2005b: 23)

and were likely to identify with the label backpacker, and periods when they were working and 'living' in a unit similar to their life at home. 'We were living in a unit for five months and we were working full-time jobs, so I didn't really feel like a backpacker. I just felt like someone who was working abroad' (Manly Backpackers Focus Group, 2006). 'I feel like I'm at home, but I happen to be have less clothes' and '[m]ost of us are here on working holiday visas, so we're sort of backpacking round for a certain amount of time, and then for the rest of it trying to work for funds to travel around' (Bondi Backpackers Focus Group, 2006). They feel 'at home abroad... It feels like being back home, working everyday... Not backpacking' (Manly backpacker Interview, 2006).

One German female traveller interviewed for the Sydney research captures the sense of ambiguity (including of resident versus traveller) as follows: '...when I came here I did get the work permit, and although I was living in a luxury apartment with my friend and worked in a really good bar, I still considered myself to be a backpacker since all those luxury apartments are actually full of backpackers. I don't think we have any residents there at all' (Pyrmont backpacker Interview, 2007). Whether the additional funds enabled by employment and the use of rented accommodation qualifies these travellers to be called 'flashpackers', however, is by no means clear. This slippery movement of travellers, in and out of work and leisure, is perhaps one of the defining characteristics of the contemporary back-cum-flashpacker in the Australian context.

Content analysis of transcripts and surveys of the Sydney backpacker research identified areas of agreement and divergence on issues that require management strategies. To navigate the precarious space, we are trying to elicit in this chapter between 'visitor', 'tourist', 'backpacker', 'working holiday maker', 'temporary' and 'permanent' resident, and the 'local' versus the 'other'. The official definitions of a backpacker and the associated regulations that form the basis of related development control plans, visa regulations, accommodation guidelines and regulations, and various Council strategies to both attract and to manage the backpacker sector (backpackers themselves and the industry) currently tend to assume the difference between each of these categories is straight forward. Yet, as one Council staff member from the City of Sydney put it: '... in many ways, given the transience of city residents, many backpackers are actually here longer than some city residents' (Backpacker Operators Focus Group, 2006). The label ascribed to groups not only conjures certain attitudes and assumes certain behaviours, it also determines legal categories – determining and triggering myriad regulatory controls. For example, the Sydney research made recommendations regarding the (mis)match between some policies and regulations and desired management outcomes. Not least, the intent of the WHM scheme to encourage long-stay visitors is at odds with the limitations regarding tenancy laws and residency in official backpacker accommodation.

The use of the common vernacular 'backpacker' can be traced to the 1970s, with the attempt by Erik Cohen (1973) to define and characterise this group of travellers (Cohen, 2003). Cohen described what he then termed the 'drifter' – invoking a distinction between institutionalised mass tourism and non-institutionalised travellers. The latter he describes as 'nomads from affluence', distinguished by the lack of fixed itinerary, a limited budget and belonging to the counter-culture intent on the 'loosening of ties and obligations, the abandonment of accepted standards and conventional ways of life, the voluntary abnegation of the comforts of modern technological society, and the search for sensual and emotional experiences.... [that motivates them] to travel and live among different and more "primitive" surroundings' (Cohen, 1973: 93). More recently, as Wilde (2006) notes, backpacker travellers have been described as 'likely to be middle class, at a juncture in life, college education and [anything but] aimless drifters. [While] they travel under flexible timetables and itineraries, most expect to rejoin the work force in the society they left' (Riley, 1988: 326). Today the 'flashpacker' – a neologism referring to apparently more affluent backpackers – is frequently just taking a break from work. Is it their budget, the accommodation they choose or the things they carry in their backpack (or other luggage given this is also a negotiable part of the defining characteristic) that differentiates them from other backpackers? Today's backpacker or flashpacker is quite likely to have a laptop, a digital camera and a mobile phone. These and other electronic devices didn't exist in the 1970s, but nor did the culture of wanting to constantly 'stay in touch', with Cohen's 'drifters' more intent on 'dropping out'. So what defines the new breed?

Cohen also noted the paradoxical gap between backpackers' ideology of 'being off the beaten track' and the reality of their practice with 'fixed travelling patterns, established routines and a system of tourist facilities and services' (Cohen, 1973: 95). As the backpacking industry has become more formalised, 'tourists' and 'backpackers' are drawn to the same destinations and sights (Wilde, 2006). The backpacker bible, the guide book *Lonely Planet*, today includes sections on accommodation from budget through to 'top end'. Perhaps the backpacker 'hippies' of the 1970s have joined the middle class, clinging nostalgically to relics of their past, while savvy entrepreneurs simply fine tune their products to meet new market demands. Are budget 'backpackers' and top end 'flashpackers' all now just versions of conventional or mass tourism? 'We are not "travellers". We're not even backpackers. We're package tourists with differently shaped luggage.... Our guidebooks lead us along a backpacking superhighway where lodgings are always cheap, natives always speak English and restaurants always serve banana pancakes' (Marshall, 1999: 58, cited in Wilde, 2006).

So who are these transient bodies and by what name should we call them? Only a small proportion (28%) of respondents to Richards and Wilson's (2004) global nomad survey identified themselves as 'backpacker'. Most considered themselves to be a 'traveller' or 'tourist' (Richards & Wilson, 2004: 144). From the 2003 International Visitor Survey in Australia, 467,985 international visitors stayed in backpacker accommodation, thus fitting within the classification of a backpacker (Ipalawatte, 2003). The same year, only 264,899 international visitors to Australia – just over half of the total number who qualified – classified themselves as backpackers.

Likewise, in the Backpackers in Global Sydney study, only 29% of respondents completing the backpacker survey selected the backpacker classification to describe themselves. Thirty percent used the category 'working holiday maker' (despite twice the number travelling on such an official permit) and 18% described themselves as 'independent traveller'. The remainder variously chose 'budget traveller', 'student traveller' or 'tourist'. Older respondents (30–35 years) were the most likely to refer to themselves as 'independent traveller' and were significantly less likely to consider themselves as 'backpackers'. But the picture is quite complicated, with as many as 30% of those over 36 years opting for the 'backpacker' label. The youngest respondents were the most likely to use the term, but still, 30% preferred to describe themselves as 'working holiday maker'. 'Backpacker', a cultural term, clearly changes meaning both temporally and spatially. In the past, 'backpacker' was a term of affection – worn as a badge of honour, to 'go backpacking' was cool. Today many who generally fit such a categorisation reject the label (Richards & Wilson, 2004) because of its negative and outmoded connotations, though once again the imprecision of the category, and the proliferation of self-definitions, appear the abiding reality from the Sydney research.

## Clashing Cultures

Whilst backpacking is the highest yielding tourism market segment in Australia and the form with the greatest probability of contributing to local economies in ways that minimise leakage, most local tourism plans state they would prefer older, quieter visitors (Bushell, 2001), who empty their wallets but nothing else. Those communities (such as Bondi in Sydney) with the resources that attract flashpackers and backpackers as long-stay travellers – notably a good supply of short- and medium-term rental accommodation, good beaches, late night pubs and clubs – have become cynical about the benefits of tourism. The discourse within many of these communities is prejudicial toward tourism, as indicated earlier in the chapter.

Consistent with the thrust of this chapter on the intensifying complexity of the backpacker phenomenon, however, the major finding of the Sydney research Residents Survey was variability, rather than consistency, of response. A majority of respondents – though not an overwhelmingly convincing one, at 47.2% – reported that backpacker tourism is an 'asset' to Sydney. Another 35% deemed it a 'liability' (with 7% reporting 'no opinion' and 11% stating 'other'). Furthermore, the Residents Survey revealed a high degree of spatial variability of opinion across the different local government areas under study. And although the responses from residents in some suburbs surveyed were far more positive than others, the split was not predictable. In some cases, suburbs with very large annual visitation numbers including a high percentage of backpackers, such as Manly, were comparatively quite supportive, and whilst concerned about certain issues, could also acknowledge the benefits. For example, only 21% of Manly respondents felt there were 'no benefits' associated with backpacking, and as many as 64% claimed backpackers 'contribute to the local economy'. Other residents, such as those surveyed in Bondi, where the tourism profile is similar to Manly and the attractions of a coastal location the same, were far less charitable. For example, as many as 42% of respondents from Waverley, the local government area in which Bondi Beach is situated, claimed there were 'no benefits' associated with backpacking. So too, in suburbs with fewer visitors of any sort, such as North Sydney, 53% of respondents reported 'no benefits'. Rather, these respondents saw only negatives. The problems attributed to backpackers included: excessive noise, binge drinking and associated late night antisocial behaviours, including large groups of 'disruptive youth' at beaches and other public spaces, vandalism, abandoned cars, illegal parking, dumped rubbish, and an incompatible degree of 'cultural difference' – all contributing to the loss of local amenity and public concern over the presence of overcrowded, unsafe hostels.

For all that, it is clear Sydney is attracting a growing number of cashed-up working holiday makers who 'are a lot more affluent' and 'don't mind paying a bit extra to get individual rooms with an ensuite' (Councils Staff Focus Group, 2006), the results demonstrate that affluence (and a desire for privacy) are not necessarily, or always, endearing the travellers to Sydney residents. Anti-social behaviour and noise continue to be significant grievances. In all local government areas in the study – except Manly – a majority of local respondents selected 'anti-social behaviour' as a problem, while noise was considered to be a problem for an overwhelming majority of respondents, especially in Waverley. As mentioned earlier, 33% of the 309 surveyed backpackers came from the UK, a further 15% from Ireland and 27% from continental Europe, the majority being male and staying in hostel accommodation at the time of the survey. This cohort, which tends to conform to the stereotypical characteristics of the more conventional backpacker, continues to elicit much antagonism, especially during the warmer months from September to March. In the words of one from the Waverley Residents' focus group (March 2006): 'I must say, yet again, its really the Irish ones who seem to be... to come here... where they come from its very rainy and cold, and when they arrive here – to have these warm evenings and all that, they go mad! (laughs)'. Even backpackers themselves, such as the German traveller mentioned earlier staying in rental accommodation in Pyrmont, were capable of finger-pointing 'the Irish and English people... who, like, without fail... are the only backpackers who are really rudely drinking... their parties were so ridiculously disgusting that we threw them out' (Pyrmont backpacker Interview, 2007).

The growth of unauthorised budget accommodation, in unregulated and sometimes dangerous premises, has intensified these allegations. The claims centre on young and highly mobile 'budget travellers' choosing to stay in non-budget accommodation and subletting to friends, or even complete strangers. This gives rise to issues of overcrowding; unreasonable noise, both in level and time of day; increased public health, fire and safety risk; increased wear and tear or damage to lifts, swimming pools, gyms and other common property items; non-adherence to bylaws in strata title buildings, including the consumption of alcohol in common areas; and the loss of security that is thought to arise from the volume of transient persons in residential premises (Allon *et al.*, 2008b). 'It's like an open house. Two-bedroom flats, so you're looking at sixty people within these four units, and there's a party in at least one room, so if they come over and they, want to have a drink and a party, and so... it's Party Central, basically' (Coogee Backpackers Focus Group, 2006). Backpacker operators estimate that there are 600–700 legal beds in Bondi, but between 1500 and 2000 illegal beds. As one interviewee put it: 'Council officers are aware of instances of owners being unaware that their units are illegal accommodation let by unethical or unlicensed agents. On other occasions it may be the landlords that are seeking to maximise the rent in their unit without regard for their neighbours' amenity' (Waverley Council Interview, 2006).

Across Sydney, local government is grappling with what has become known as the 'backpacker phenomenon'. Places and localities within them are increasingly becoming 'contact zones', criss-crossed and traversed by flows and circulations of people, capital (social, cultural and economic), objects and ideas (Pratt, 1992). People from different places and world views, with radically different purposes within the same place, are being juxtaposed and connected in ways not experienced by previous generations (Delanty, 2003). In this age of globalisation, with the democratisation of information through cheap and pervasive enabling technologies, these mobile people have at their disposal sophisticated, informal networks that transcend the jurisdiction and control of local government bodies. The encounters between tourists and tourist destinations cannot, therefore, be understood simply as 'visitors' interacting with stable, permanent and 'authentic' localities or local communities.

A locality like Bondi in Sydney, for example, is increasingly dependent on the labour and service of backpackers who sustain the cafes, restaurants, surf shops, grocery and other stores that are valued by residents and visitors alike (Waverley Chamber of Commerce Focus Group, 2006). With accelerating mobility and intensifying connectivity, tourists and destinations, guests and hosts, are reciprocally transforming. In Global Sydney, backpackers are one particular inflection of a globalisation process in which the concept of the 'local' is thrown into question. In a suburb like Bondi Beach, 'outsiders' are met with varying degrees of welcome, acceptance, resistance and hostility. 'Tourism NSW and Australian Tourism is saying one thing, and poor Bondi and Waverley are trying to say, "It's not like that, guys, because we can't handle it"' (Waverley Resident Focus Group, 2006).

The Backpackers World Travel website advertising on behalf of the Fiji Visitor Board agrees with Tourism Australia and the residents of Bondi, that backpackers are actually easily spotted. In this case, however, the Fijian government is positioning the backpacker as a more desirable cultural tourist, the advertisement stating: 'they are the ones in departure lounges without a cheesy lei around their neck, they are not sipping artificial fruit juices and their travel stories are very different to the ones who simply visited an island resort. These travellers are the ones leaving with a real understanding of the Fijian culture' – an apparently more 'authentic' experience because they have traversed the boundary of the visitor. The question from the locals' perspective is whether this is desirable, or whether they prefer it when the visitor remains within a membrane that preserves the boundary between the local and the visitor? Or is the more pertinent question, who *are* the locals? In traditional societies, like some Fijian villages, this can perhaps be readily answered, but in Australia's largest city and Pacific Rim gateway, this is very complex. Will newer forms of long-stay traveller, including the so-called 'flashpacker', be more responsible and acceptable from the perspective of the 'local'? How are high-impact tourist localities best managed, and what governance and planning policies best serve places where the line between the local and the visitor is becoming increasingly porous? It is at the juncture between dwelling and travelling, and between mobility and stasis, where conflicts emerge over the identities of places and the production of localities as tourist commodities.

**Note**

1.   In March 2007, the Advertising Standards Authority in the UK ordered the removal of billboards with the slogan. The advertisement was also banned by regulators in Canada and the USA (*Sydney Morning Herald*, 2007). In Singapore, the campaign was re-presented as 'So Where Are You?' with 'bloody hell' removed.

## Chapter 14

# Towards Strategic Planning for an Emerging Backpacker Tourism Destination: The South African Experience

CHRISTIAN ROGERSON

## Introduction

South Africa is a latecomer to the competitive international business of backpacker tourism (Rogerson 2007a, 2007b). The local industry of backpacker tourism was undeveloped as a consequence of global sanctions and associated tourist boycotts of apartheid South Africa. Since the 1994 democratic transition, however, backpacker tourism has experienced a period of expansion, particularly during 'the Mandela Boom', with a surge in international tourism arrivals to South Africa. Under the direction of South African Tourism (SAT), strategic planning for the post-apartheid tourism economy initially focused upon promoting South Africa for low volume, high value, long-haul international tourism (Cornelissen, 2005). Within this planning focus, backpacking was a neglected element of the tourism economy (Visser, 2004, 2005; Visser & Barker, 2004). After 2006, however, national tourism planning shifted to recognise the strategic significance for South Africa of promoting youth and budget tourism, including a focus on backpacker tourism.

The aim in this chapter is to unpack the development of policy and planning for backpacker tourism in South Africa. In terms of backpacker studies, this contribution is situated as part of a wider body of applied research that is explicitly linked to policy issues around backpacking (e.g. Hampton, 1998, 2003). More specifically, this research links to what Richards and Wilson (2004: 10) describe as 'helping practitioners deal with the practical consequences of the expansion of backpacker tourism'. A tradition of strategy-focused works concerning backpacker tourism is most strongly developed within Australian and New Zealand based research (see e.g. Mohsin & Ryan, 2003; Ipalawatte, 2004; Newlands, 2004, 2006; Prideaux & Coghlan, 2006b; Prideaux & Shiga, 2007; Trembath, 2008). In terms of source material, the chapter mainly draws from two sets of material: (1) a wealth of policy documentation on the South African tourism economy as a whole; and (2) material collected from a 2006 national study on backpacker tourism in South Africa, which was conducted for the purpose of developing a national strategy for backpacker tourism. This research involved undertaking 125 interviews with suppliers of backpacker tourism products in South Africa and 300 consumer interviews with international backpackers.

The analysis is divided into three sections. First, a brief context is given of tourism growth in South Africa and the development of tourism policy, especially since the 1994 democratic transition. Second, a profile of backpacker tourism in South Africa is presented. Third, key findings relating to the constraints or barriers to competitiveness on backpacker tourism in South Africa are highlighted. By way of conclusion, current directions and progress of policy support towards backpacker tourism in this emerging backpacker destination are discussed. Overall, this study contributes to the limited existing scholarship around policy and planning for the promotion of backpacker tourism within a developing world country.

## Tourism and Tourism Policy in South Africa Especially Post-1994

Until the democratic transition, the expansion of international tourism as a national policy priority in South Africa was neglected (Rogerson & Visser, 2004). This situation is hardly surprising as the country's mass tourism industry is barely 40 years old and anchored in the domestic market. In 1947, a special government agency, South African Tourist Corporation (SATOUR) was established and commenced promoting the country as a tourism destination through several branch offices in Europe, North America and Australia. Nonetheless, even by the mid-1960s, tourism in South Africa remained a local affair with the country's immediate neighbours providing the mainstay of the 'international' tourist intake. Four out of every five 'international' visitors were either white residents coming from the colonial territories of Southern or Northern Rhodesia, spending their vacation periods on the beaches of Durban and Cape Town, or Portuguese visitors from colonial Mozambique travelling to Johannesburg for entertainment or shopping. Long-haul tourists represented only 17% of tourist arrivals, usually British visitors, often arriving in South Africa after a two-week sea journey (Ferrario, 1978).

Although international air travel to South Africa escalated from 1957, a hindrance to the growth of the national tourism industry was the high cost of international air fares, especially from Western Europe. From 1966, international tourism was spurred variously by offers of low-cost excursion fares to south-bound air traffic between Europe and South Africa, the progressive introduction of larger jet aircraft and increases in scheduled international flights. Official policy, however, was to discourage the development of mass international tourism through cut-rate charters, which might result in the inflow of 'hundreds and thousands of comparatively penniless and permissive tourists who would overtax the available facilities and pose a threat to the traditionally conservative outlook of the South African way of life' (Ferrario, 1978: 63). South African tourism marketing targeted a high income, low volume tourist intake and the branding of the country as 'a sophisticated, rather select destination, appealing to what can be termed the inflation-proof section of travel markets in the northern hemisphere' (SATOUR, 1975: 7).

Despite the growing implementation of apartheid policies, international tourism steadily expanded throughout the 1970s (Figure 14.1). The 1976 Soweto uprising marked the beginnings of a new phase of political isolation with the introduction of economic sanctions and the consolidation of South Africa as an international pariah state (Rogerson & Visser, 2004). The isolationism of the old apartheid system delayed South Africa's entry on the global tourism stage, an entry that witnessed a

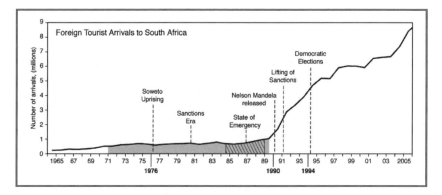

**Figure 14.1** Trends in international tourism 1965–2006
(Source: South African Tourism Annual Report, 2006)

spectacular expansion of international tourism arrivals after the release of Nelson Mandela (Figure 14.1).

Building upon the 'Mandela boom' and the lifting of economic sanctions, an enormous expansion in the country's tourism occurred. This period of strong growth in tourism arrivals since 1990 'fundamentally changed the face of the tourism industry in South Africa' (SAT, 2007a: 15). In terms of global tourism, however, even by 2006, South Africa remains a minor player representing 8 million international tourism arrivals in a global market of 800 million. But, since 1995, South Africa has been, alongside Egypt, the major international tourism destination in Africa (Rogerson, 2007c). Significantly, 73% of South Africa's international tourism arrivals are accounted for by 'regional tourists' from sub-Saharan Africa rather than long-haul international travellers. Of the top 10 country source markets for international tourists, six of South Africa's biggest markets – Lesotho, Swaziland, Botswana, Zimbabwe, Mozambique and Namibia – are neighbouring countries (Rogerson & Visser, 2006; Rogerson & Kiambo, 2007).

Cassim (1993) argues that the period of the early 1990s represents a crucial watershed, for it is at this important juncture of South Africa's development history that tourism begins to seriously enter the realm of policy debate. For Lowitt (2006: 18) 'the allure of a strong tourism sector for South Africa is strong'. Since 1994, new policy frameworks have emerged to support the development and changed role of tourism (Rogerson & Visser, 2004). Institutionally, tourism is the core line function of the Department of Environmental Affairs and Tourism (DEAT), but importantly, is also a concurrent function of the Department of Trade and Industry (DTI), which is responsible for promoting national economic growth. The key organisation for policy implementation is South African Tourism (SAT). Under the Tourism Act, SAT's mandate is to deliver sustainable GDP growth, sustainable job creation and promote redistribution through six key objectives, namely, increase tourism volume; increase tourist spend; increase length of stay; improve geographic spread; improve seasonality; and promote 'transformation', which refers to changing the racial complexion of ownership and beneficiaries of the tourism economy (Rogerson, 2008a).

By 2006, tourism was a major contributor to the South African economy at an estimated 8.3% of GDP, 947,530 jobs or 8% of total employment (DEAT, 2008: 28). Boosted by South Africa's imminent hosting of the 2010 FIFA World Cup, Government targets increasing foreign tourist arrivals to 10 million and an additional 400,000 direct and indirect jobs in tourism by 2014 (DTI, 2007a: 21). The critical long-term importance of tourism for the South African economy is highlighted by the 2006 Accelerated and Shared Growth Initiative for South Africa (ASGISA), which represents a government-driven economic development programme (DTI, 2005a). In line with national objectives seeking to halve poverty and unemployment by 2014, ASGISA aims to energise South Africa on a pathway of higher economic growth. But ASGISA's emphasis is not just on the urgency for growth but instead for achieving a wider spread of benefits of accelerated growth, or, in other words, shared growth (DTI, 2005a).

Under ASGISA, a group of priority sectors are identified that represent 'low-hanging fruit' deemed ripe for further development (DTI, 2005b; Rogerson & Visser, 2006). The targeted 'priority sectors' are viewed as those sectors in which the country has comparative economic advantages, which, if fully exploited, would lend themselves to higher rates of economic growth (DTI, 2005a). Tourism is one of the immediate priority sectors in which it is considered support programmes can be implemented (DTI, 2005b). More specifically, tourism is considered a potential means of fostering 'shared growth', in that it has potential for economic growth as well as features that can benefit the poor. Tourism is rated highly in terms of its potential for poverty alleviation, not least through the creation of much needed employment opportunities in the country's poorer regions and the interventions of pro-poor tourism projects (Ashley & Haysom, 2006; Rogerson, 2006). Cornelissen (2005: 42) argues that tourism in South Africa is regarded 'as a key catalyst for the economic growth that the government would like to attain to meet the country's developmental imperatives'.

Traditionally, for international tourists, South Africa 'is an exotic destination that offers the tourist the possibility to become an explorer and in many ways to refashion a romantic pioneering era' (Cornelissen, 2005: 55). After democratic transition, the South African tourism product initially centred upon the attractions of landscape, wildlife and culture. Indeed, in common with the apartheid era, the country continued to avoid any marketing for mass-packaged beach tourism, a market that has only been rudimentarily developed in South Africa (Cornelissen, 2005: 55). For the large volume of regional tourists from sub-Saharan Africa, shopping and business are major attractions (Rogerson, 2004; Rogerson & Kiambo, 2007). Recent international marketing of South Africa is guided by the priorities articulated in SAT's Growth Strategy, first developed in 2002 and revised on two subsequent occasions. The Tourism Growth Strategy focused national marketing initiatives upon a core portfolio of countries selected for their relative importance and attractiveness to

South Africa (SAT, 2004: 9). Within these key consumer markets, market segmentation has taken place in order to identify attractive consumer segments (SAT, 2004: 11). Significantly, the first national Tourism Growth Strategy of 2002 (SAT, 2002) excluded backpacker tourism from the 'growth' segments of tourism that were to be targeted for priority attention (Visser, 2004, 2005). Instead, the main focus continued to be marketing South Africa to high value, long-haul travellers seeking an 'exotic' safari destination.

By 2007, a marked policy change is evident in the appearance of the third Revised Tourism Growth Strategy for 2008–2010 (SAT, 2007b). Although this strategy retains and consolidates early visions, it argues that despite an eightfold increase in tourism arrivals since 1991, 'further growth is essential if tourism is to make a meaningful impact on job creation and GDP growth' (SAT, 2007b: 15). Importantly, it stresses that 'South Africa cannot be a high end low impact destination for a few wealthy foreigners'; rather, there is a need to focus on growing volume and reaching beyond the exclusive five and four star markets (SAT, 2007b: 15). By the mid-2000s, it was apparent that over-capitalisation and excess capacity existed in some parts of the tourism economy, most notably in the segment of luxury game lodges (DTI, 2005b: 13). Further, SAT (2007b) concedes that in the minds of international consumers, South Africa remains 'much the same as what it was 10 to 15 years ago'. It was acknowledged that 'South Africa is still perceived mainly as an adventurous wildlife destination with striking natural beauty'. Among the several changes introduced in the 2007 strategy is the objective of opening the country 'to younger travellers in whom we could make a lifetime investment as potential repeat travellers in South Africa at different stages in their life' (SAT, 2007b: 15). It is against this backdrop of changing national tourism policy that attention turns to examine key facets and policy challenges of South Africa as an emerging backpacker destination.

## Key Features of an Emerging Backpacker Destination

Before 1994, the backpacking industry of South Africa was undeveloped, suffering from the effects of sanctions on international tourism owing to the country's apartheid policies. Since the 1994 democratic transition, South Africa has emerged as an increasingly popular backpacker destination seeking to compete in the international economy of backpacker tourism (Visser & Barker, 2004; Visser, 2005; Rogerson, 2007a, 2007b; Niggel & Benson, 2008). This section summarises key findings from the recent national investigation of backpacker tourism in South Africa. The backpacking industry in South Africa is viewed here as comprising three sub-segments of enterprises that are concerned, respectively, with the supply of (1) accommodation, (2) transportation and tour services and (3) adventure tour products.

It has been estimated that by 1994–1995 there were less than 50 established backpacker enterprises in operation across the country. Since the 1990s, development and expansion of a network of backpacker enterprises has taken place (Rogerson, 2007a). By 2006, the industry is constituted by a total of 500–600 formal enterprises as well as an unknown number of unregistered or 'informal sector' enterprises. Largely, the industry is the domain of small tourism firms with the average backpacker enterprise employing seven workers. The backpacking economy contains a large cohort of family-owned enterprises, many of which have been established by 'developer-tourists'. The industry burgeoned as entrepreneurs identified opportunities for business development subsequent to political change and South Africa's re-entry into the international tourism economy. The institutionalisation of the industry is reflected in the appearance of Backpacking South Africa, an association of suppliers of backpacker products, and by *Coast to Coast*, a guidebook to backpacking in Southern Africa. A significant factor in the institutionalisation and growth of a backpacker network in South Africa has been the activities of the dedicated backpacker transportation and tour enterprise (Baz Bus), which is an example of developer-tourist enterprise founded in 1995 by a former qualified accountant and retired backpacker. Many suppliers of backpacker accommodation themselves are former backpackers and qualify as developer-tourists. A segment of these entrepreneurs would also be classed as 'lifestyle entrepreneurs', choosing for personal motivations to establish backpacker businesses often in attractive localities.

In racial terms, the South African backpacker industry is distinguished by its domination by white entrepreneurs. Notwithstanding national government initiatives for the transformation of the national tourism industry, only a handful of individual black entrepreneurs have entered the backpacking industry. The most popular avenue has been through establishing backpacker enterprises in new market spaces, such as urban townships. The best example is Lebo's backpackers in Soweto, a pioneer backpacker accommodation establishment that has tapped into the international market for 'township tourism' (Mograbi, 2007). Another channel of black empowerment is the recent establishment, with government-funding support, of several community-owned backpacker tourism enterprises. In Eastern Cape Province, community equity partnerships have been formed with existing (white-owned) backpacker enterprises. In terms of overall patterns of ownership, the national survey of 125 backpacker product suppliers found only four enterprises owned by black individuals; in addition, the survey captured one community-owned backpacker enterprise, and three enterprises with community-equity share arrangements.

Figure 14.2 shows that the South African backpacking industry, as indexed by numbers of enterprises, is overwhelmingly focused in Western Cape Province (centred around Cape Town) followed by the provinces of Eastern Cape, KwaZulu-Natal, Gauteng and Mpumalanga. The greatest share of enterprises in the industry is accounted for by suppliers of accommodation for backpackers, which range from boutique-style backpacker accommodation as offered in Cape Town, to simple dormitory or *rondavel* accommodation as found in most rural areas. The largest clusters of backpacker accommodation providers occur

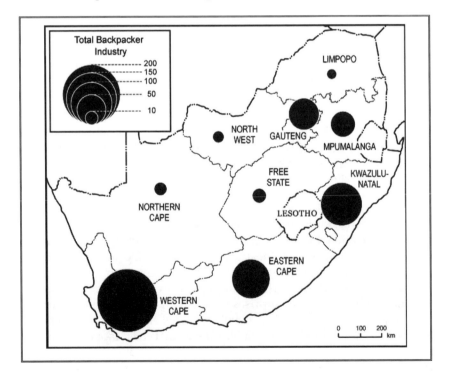

**Figure 14.2** The spatial distribution of enterprises involved in backpacking

in Western Cape Province followed by the Eastern Cape, KwaZulu-Natal, Gauteng and Mpumalanga. Only limited provision of backpacker accommodation occurs in the four provinces of Free State, Northern Cape, North West and Limpopo (Rogerson, 2007a). Not surprisingly, the current patterns of visitation and circuits of travel of international backpacker tourists are geographically uneven. The national survey confirms the role of the Western Cape and more especially of Cape Town as the outstanding focal point for international backpackers in South Africa (Visser & Barker, 2004b; Rogerson, 2007c). On average, 75% or more of international backpackers that visit South Africa stay at least one night in Cape Town.

Figure 14.3 reveals the 20 most important South African 'hotspots' for international backpackers as ranked by the proportion of backpackers in

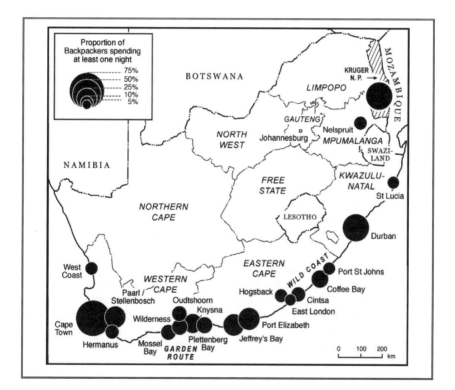

**Figure 14.3** The spatial distribution of the backpacker visits
(Note: Johannesburg is excluded from the analysis as the base for the interviews)

the consumer survey that spent at least one night there. Based upon exit interviews conducted in Johannesburg with 300 international backpackers, Figure 14.3 indicates the channelling of backpackers along the coastal route from Cape Town to Port Elizabeth, the significant destinations of the Wild Coast, the importance of Durban and the wildlife attractions of Kruger National Park. Cape Town and Johannesburg are the major gateways for backpackers into South Africa and Johannesburg (OR Tambo International) the core port of exit (Mograbi, 2007). As a whole, the provinces most frequented by international backpackers are Western Cape, Eastern Cape, KwaZulu-Natal and Mpumalanga; least visited are Free State, Limpopo, North West and Northern Cape. At a provincial level, international backpackers travel on average to 3.84 of South Africa's nine provinces, which compares favourably to the average international tourist who visits only 1.41 provinces (SAT, 2007a). Accordingly, in common with other international findings (Hampton, 1998), backpacker tourists to South Africa contribute to a wider geographical dispersal of the benefits of a growing tourism economy than do groups of higher spend international visitors to the country.

Official data on the size of the backpacker tourism market in South Africa is unavailable because the category of 'backpackers' has not been included in regular SAT surveys of international and domestic tourism. It is evident that since democratic transition, a rapid growth has occurred from an almost negligible base during the early 1990s. Using existing SAT information on bed nights, the best estimate is that the South African backpacking industry (2006) attracts 90,000 international backpacker visitors, who spend at least one night in backpacker/hostel accommodation per trip (DTI, 2007b). In comparative terms, this figure should be set against Australian data (2005), which indicates 482,000 international backpackers and that backpackers represent approximately 10% of Australia's international visitors. In South Africa, the share of backpackers in overall international tourism arrivals presently represents less than 1%, a signal of considerable growth potential for the future.

The country's backpacking industry is overwhelmingly reliant upon international rather than domestic tourists. This trend was confirmed in the 2006 national study, which found that 96% of interviewees were international backpackers (Rogerson, 2007b). From parallel product surveys of backpackers, the market reliance upon international backpackers was again reinforced (Mograbi, 2007; Rogerson, 2007a). The size of the domestic backpacking industry is estimated at only 22,000 visitors, approximately a quarter of the size of the international backpacker numbers. The underdevelopment of domestic backpacking is a distinctive feature of the backpacking industry in South Africa, particularly as compared to Australia, which records total numbers of domestic back-packers comparable to those of international backpackers.

The key features of backpacker tourists were disclosed from the findings of the national consumer survey. In terms of the origin or source of international backpacker tourists to South Africa, 22 different nationalities were captured in the survey. Nevertheless, the major sources for backpacking in South Africa are highly concentrated. More than 80% of the South African backpacker market is represented by only six countries, namely, the UK, Australia, the USA, Germany, New Zealand and the Netherlands (Mograbi, 2007; Rogerson, 2007b). Amongst these leading six source countries, a high proportion of visitors from the USA were found as part of volunteer tourism programmes and engaged in community or social work, medicine and conservation.

The average expenditure in South Africa of international backpackers is an estimated total of R10,294 per trip (excluding the long-stay volunteers) or a total of R13,395 per trip, which includes volunteers (US$1 = R7.20 at 2006 exchange rates). Although the average international backpacker spends 42 nights in the country, 72% of backpackers stay in the country for a month or less. The decision to backpack in South Africa was exclusive to only 56% of international backpackers; the rest visit South Africa as part of a wider international travel itinerary. Significantly, however, the length of stay of international backpackers in South Africa was found to be less than that allocated to visits in other competing backpacker destinations, such as South East Asia (average 2 months), Australia or New Zealand (3–6 months or up to one year), South America (2–3 months) or the UK (9 months). A substantial segment of backpackers are not staying as long a period of time in the country as they might wish to, in large measure because of their lack of awareness and access to information about the country's tourism products (Mograbi, 2007; Rogerson, 2007b). An added factor is the absence in South Africa of opportunities for working holidays, in marked contrast to Australia or New Zealand, which offer working holiday visas to international backpackers (Ipalawatte, 2004; Newlands, 2004, 2006). The longest periods of stay by international backpackers in South Africa are recorded by volunteer tourists, which represent one of the fastest expanding segments of the broader industry of youth tourism in South Africa.

The findings of the consumer surveys reveal a generally positive experience of international backpackers concerning their visit to South Africa and of the country's tourism products (Rogerson, 2007b). Several important advantages were highlighted of the South African backpacker experience. The most frequently cited advantages concerned the country's natural attractions, especially the wildlife and game parks, and the wealth and diversity of activities on offer, from sightseeing to extreme adventure sports. For many international backpackers, culture was another important positive element of the South African backpacker experience, with issues of the country's political history as well as African culture flagged as attractive. Despite complaints about the costs and inflexibility of travelling with Baz Bus, international backpackers rated South Africa as a country that was 'easy to travel', with many opting to travel by hire car. The well-organised nature of local backpacking and the high quality of services, cleanliness and facilities at the country's backpacker hostels were given high ratings. Further, attention was drawn by backpackers with experience of travel in other parts of Africa, South America or Southeast Asia to the comparatively excellent standards of infrastructure across South Africa.

The consumer surveys disclosed that whilst the overwhelming majority of international backpackers would recommend South Africa as a backpacker tourism destination, their experience of backpacking in South Africa 'was not really backpacking' (Rogerson, 2007b). This critical observation arose out of the reduced ability for freedom of independent travel in South Africa due to the combined effects of poor public transportation and issues surrounding crime and safety that are endemic problems of the local tourism industry (Rogerson & Visser, 2004). The combined impact of these factors is to limit the potential for 'independent travel' that is identified at the heart of backpacking as a distinctive form of tourism (cf. Pearce, 1990, 2006).

## Strategic Challenges Facing Backpacker Tourism

In terms of strategic challenges facing the development of backpacker tourism in South Africa, it should be acknowledged at the outset that at one level these challenges are inseparable from the constraints confronting the development of South African tourism as a whole (SAT, 2004, 2007b). The core problems that must be addressed in terms of enhancing South Africa as a competitive international destination are widely acknowledged as relating to inadequate airlift capacity and aviation pricing; poor public transport; crime, safety and tourist security and the general perception of South Africa as an 'unsafe' destination; the branding of South Africa, insufficient marketing and provision of relevant information; the imperative for innovation and new product development in the country's tourism economy; and improving human resources and skills in tourism (Rogerson & Visser, 2004; SAT, 2004, 2007b; DTI, 2005b; DEAT, 2008).

From the results of the national investigation on backpacking, a number of more specific barriers to the competitiveness of the local backpacking industry were identified. Fifteen key barriers or constraints to competitiveness were discerned.

(1) The backpacking industry lacks official recognition by all levels of government as a viable sector of South African tourism.

The enterprise surveys clearly revealed that the backpacking industry operates in an environment in which there was little recognition of its potential as a valuable segment of the national tourism economy. In addition to a lack of recognition by national government, at sub-national levels of government there was minimal recognition of the value of the backpacking industry.

(2) The backpacking industry as part of 'budget tourism' as a whole is a misunderstood segment of tourism and suffers from a poor image in South Africa.

The backpacking industry in South Africa suffers from a relatively poor image as part of the wider category of 'budget tourism'. There existed widespread lack of awareness and understanding of the distinctive features of backpacker tourism at all levels of government, as well as in provincial and local tourism associations, banks and other support organisations.

(3) The South African backpacking industry is inadequately marketed to compete with a range of existing and emerging international destinations.

The inadequacy of international marketing of South Africa as a backpacker destination was identified by product suppliers as the key challenge facing the industry. In an increasingly competitive international backpacking economy, with aggressive marketing of other destinations (especially Australia and New Zealand), the potential for industry development was reduced by limited marketing of South Africa to backpackers.

(4) The backpacking industry is negatively affected by the limited working holiday visas programme that operates in South Africa.

One special factor that negatively impacts upon South Africa's competitiveness in the international backpacker economy is the limited opportunities for working holidays. It was disclosed that whilst nearly three-quarters of international backpackers interviewed would be interested in opportunities for undertaking a working holiday in South

Africa, only a tiny proportion of backpackers currently undertake any work during their stay (Rogerson, 2007b). Since 2000, the Department of Home Affairs has granted South African Student Travel Services (SASTS) the mandate to process applications from 35 international partner organisations for a working visa programme, allowing qualified applicants (aged between 18 and 25, graduate or full-time university student) to obtain a 12-month visa issued overseas by South African embassies or High Commissions. However, the current scale of this programme is running at only 30–40 successful applications per year; in certain years, Home Affairs intervened to block the programme entirely. The existing costs and procedures of accessing the work visa through this programme are burdensome as compared to the ease of access of application for a working holiday visa in either Australia or New Zealand.

(5) The backpacking industry is negatively affected by inadequate awareness of volunteer placement opportunities and of existing local programmes for educational exchange and study abroad.

The consumer survey disclosed the importance of volunteer tourists in the current market for international backpackers in South Africa. Survey respondents highlighted a need for improved awareness of opportunities for volunteer work placements in the country. In addition, as compared to the aggressive marketing of short-term study-abroad opportunities in Australian tertiary institutions, there was only limited or weak marketing of such opportunities for foreign students at South African tertiary institutions.

(6) The backpacking industry is constrained by the weakness of the domestic backpacking sector.

In comparative international perspectives, one of the most distinguishing features of the backpacking industry in South Africa is the limited number of domestic backpackers. The consumer surveys of international backpackers reported that most met 'very few' domestic backpackers on their travels and that there was a marked disappointment among international backpackers with the limited interactions with domestic backpackers. In terms of existing market volume, the greatest volume of domestic backpackers is comprised of school, sports or church groups, rather than individual travellers. The product survey research identified that in certain parts of the country (particularly in the Western Cape), negative attitudes prevailed about the domestic backpacker market. Many suppliers of backpacker accommodation voiced complaints that 'South Africans are too much trouble' or 'bad mouth South Africa'. The 2008 Coast to Coast Guide for backpackers issues 'a word of caution to

South Africans' that 'some places mentioned in Coast to Coast only accept international travellers' (Coasting South Africa, 2008: 7).

(7) The South African backpacking industry lacked a strong and effective industry association.

The development of the product platform within the backpacking industry represents another essential dimension for enhancing competitiveness. At the industry level, one of the most critical lessons from a review of the international experience is the importance of developing strong industry associations to support and provide a policy lobby on behalf of the backpacking industry (Rogerson, 2008b). A major contrast exists between the situation in Australia, where a strong industry association in backpacking has been nurtured, and the current state of Backpacking South Africa (BSA), the parallel industry association. In Australia, the counterpart industry association – the Backpacker Tourism Advisory Panel – receives support through national government funding. By contrast, BSA was funded only through voluntary membership subscriptions, a situation that seriously limits the activities that can be realistically undertaken to promote and improve the industry.

(8) At the firm level there is a need to maintain the existing quality standards in the backpacking industry and rectify certain weaknesses in the existing product offering.

Although the findings from the consumer surveys revealed a generally positive experience of international backpackers concerning their visit to South Africa and of the country's tourism products, certain shortcomings were noted. First, international backpackers expressed a number of concerns about the inflexibility and pricing of Baz Bus, which since 1995 has enjoyed a monopoly on dedicated backpacker transportation services in South Africa. Second, in terms of activities that were undertaken by backpackers in South Africa, three attracted a degree of criticism. The activity identified as 'most disappointing' by international backpackers was canopy tours. Another group of criticisms were directed at cultural villages, which for many backpackers represented 'staged' rather than 'authentic' experiences. Lastly, in terms of negative responses, South Africa's night-life and night clubs were an added source of disappointment (Rogerson, 2007a).

(9) The backpacking industry suffers from limited access to bank finance for product upgrading and development of new products.

The upgrading of facilities and tourism products in backpacking is mostly undertaken on the basis of the entrepreneur's own risk capital or from retained earnings. Currently, the existing facilities in South African accommodation hostels are highly rated as compared to similar facilities in other parts of the world. Indeed, one popular website promoting independent youth travel observes that: 'Some would say (it's a close call with New Zealand) that South Africa has the best hostel network in the world' (Travel Independent, 2008). In order to maintain international competitiveness, this quality level must be maintained in South Africa through continued upgrading. Although access to finance is not a major constraint on the competitiveness of the backpacking industry as a whole, the product survey disclosed that the banking sector is an unwilling partner in finance for supporting backpacker enterprises, in part due to its lack of understanding of the workings of the industry. In addition, in the supply of transportation services, entry into the backpacking industry is constrained by the high capital requirements for the purchase of vehicles.

(10) The backpacking industry suffers from the limited number of black entrepreneurs who are currently involved in this industry.

The enterprise research disclosed that the backpacking industry of South Africa is currently a segment of tourism that is overwhelmingly dominated by white entrepreneurs. The competitiveness of this segment of South African tourism could be strengthened by the encouragement of a wider diversity of backpacker entrepreneurs, including expanding opportunities for community-owned enterprises or share-equity arrangements, in order to further diversify the product platform.

(11) The backpacking industry is constrained by the lack of subnational development initiatives concerning product development.

At the destination level in South Africa, the product surveys disclosed that both provincial and local development planning overlook the potential offered by the backpacking industry. The activity of Eastern Cape Province in supporting backpacking provides the only exception. Currently, local tourism officers and those involved in either local economic development (LED) or local tourism planning often lack knowledge of opportunities in backpacking.

(12) The backpacking industry is impacted negatively at times by the absence of appropriate regulatory frameworks and at other times by excessive or inappropriate regulations.

The enterprise surveys with accommodation suppliers highlighted the negative impacts on the development of backpacker enterprises because of the absence of any local government regulations that are specific to and sensitive of the needs of this segment of the tourism industry. The enterprise surveys showed that in several municipalities there are negative attitudes and outright opposition to backpacker tourism. Backpacker travel and tour suppliers point to additional problems concerning 'fly-by-night' operators and the poor quality of services offered that negatively impacts upon the backpacking industry as whole (Rogerson, 2008c). In addition, the enterprise surveys further show that in other areas of their operations relating to the transportation of tourists, there exist situations of excessive 'red tape' (concerning transport permits and licensing) or inappropriate regulations that also negatively affect the industry (Rogerson, 2008c).

(13) The backpacking industry suffers from limited regional incentives available for tourism, especially to support tourism in South Africa's peripheral regions.

The international experience of backpacking is that it can be an important vehicle for energising regional and local development in undeveloped or lagging regional economies (see Hampton, 1998, 2003; Donaldson, 2007). Regional development support in South Africa has traditionally focused on the manufacturing sector. It is essential that new regional development strategies in South Africa recognise the signifi-cance of tourism as a competitive sector for promotion and support in peripheral areas.

(14) The backpacking industry is impacted by South Africa's poor public transport provision and chronic issues of the safety of tourists.

The consumer surveys disclosed that poor public transport and perceptions of safety and crime limit the activities and freedom for the independent travel of backpackers in South Africa.

(15) The South African backpacking industry suffers from inadequate research and limited accurate statistics.

The strengths and contributions of the backpacking industry to the tourism economies of Australia and New Zealand have become evident to key stakeholders through the availability of accurate statistics and regular research updates concerning the backpacking sector (Rogerson, 2008b). By contrast, no regular monitoring of the backpacking sector has been undertaken in South Africa. Existing SAT data understates the significance of the industry and regular statistics to monitor the backpacking industry are absent.

## Conclusion

Since 2006, the policy environment towards backpacker tourism in South Africa has radically altered because of recognition of the potential importance of youth travel. The roll out of a national support programme is being coordinated through the Tourism Unit within the DTI, the line ministry responsible, in 2005, for launching a customised sector development programme for tourism as a whole (DTI, 2005b). The DTI has taken the lead in identifying various 'niche' forms of tourism in which South Africa is viewed as having a competitive advantage. Among several types of niche tourism, backpacker tourism is identified as one that offers considerable potential for future expansion in South Africa (DTI, 2007b). Australia has been used as a competitive benchmark for the future development of backpacking and in South Africa there is considerable interest in learning policy lessons from that country (Rogerson, 2008b).

During 2007–2008, the initial steps have been taken by DTI to support the South African backpacking industry. First, a greater awareness of local opportunities in backpacking has been fostered through its funding of a booklet that highlights opportunities in the industry (DTI, 2007b). This promotional booklet is to be distributed and supported through workshops with local government tourism officers, to disseminate opportunities for tourism at grassroots level and through the networks of SEDA, the national small business development agency (DTI, 2008). Second, DTI has addressed weaknesses of the industry association through channelling direct financial support to the industry association – Backpacking South Africa – for its operations and engagement of full-time staff. One signal of the greater energy of the strengthened local industry was the hosting during 2007 in Cape Town of the first national Backpacking and Adventure Conference. Third, DTI has sponsored a set of initiatives to promote domestic backpacking through national campaigns to promote backpacking as part of its Sho't Left domestic tourism promotion. Finally, and most importantly, enhanced marketing support has been offered by DTI for South Africa's backpacking industry. In particular, DTI has financed industry participation at the World Youth and Student Travel Conference (WYSTC) meeting in Istanbul, as well as encouraged bidding for hosting WYSTC in Cape Town in 2012, using that event as a platform to build beyond the 2010 football World Cup for marketing South Africa to international backpackers (Meyer, 2008). Collectively, these initiatives point to a new energy and strategic commitment on behalf of national government in South Africa in support of the competitiveness of the local backpacking industry.

# References

1770 Festival 2008 (2008) *1770 Festival 2008*. On WWW at http://www.1770fes tival.com.au/about.htm.

Adey, P. (2007) 'May I have your attention': Airport geographies of spectatorship, position, and (im)mobility. *Environment and Planning D: Society and Space* 25, 515–536.

Adkins, B. and Grant, E. (2007) Backpackers as a community of strangers: The interaction order of an online backpacker notice board. *Qualitative Sociology Review* 3 (2), 188–201.

Adler, J. (1989) Youth on the road: Reflections on the history of tramping. *Annals of Tourism Research* 12 (3), 335–354.

Aitchieson, C. (1999) New cultural geographies: The spatiality of leisure, gender and sexuality. *Leisure Studies* 18, 19–39.

Alaszewski, A. (2006) *Using Diaries for Social Research*. London: Sage.

Allon, F., Anderson, K. and Bushell, R. (2008a) Mutant mobilities: Backpacker tourism in 'global Sydney'. *Mobilities* 3 (1), 73–94.

Allon, F., Bushell, R. and Anderson, K. (2008b) *Backpackers in Global Sydney: Final Report*. Sydney: Centre for Cultural Research, UWS.

Alneng, V. (2002) What the fuck is a Vietnam?: Touristic phantasms and the popcolonization of (the) Vietnam (War). *Critique of Anthropology* 22 (4), 461–489.

Anderson, B. (1983) *Imagined Communities. Reflections on the Origin and Spread of Nationalism*. London: Verso.

Anderskov, C. (2002) *Backpacker Culture: Meaning and Identity Making Process in the Backpacker Culture among Backpackers in Central America*. Research report. Denmark: Department of Ethnography and Social Anthropology, Århus University.

Appadurai, A. (1990) Disjuncture and difference in the global cultural economy. *Theory, Culture and Society* 7, 295–310.

Appadurai, A. (1991) Global ethnoscapes: Notes and queries for a transnational anthropology. In R. Fox (ed.) *Recapturing Anthropology. Working in the Present* (pp. 191–210). Santa Fe, CA: School of American Research Press.

Appadurai, A. (1996) *Modernity at Large: Cultural Dimensions of Globalization*. Minneapolis, London: University of Minnesota.

Aramberri, J. (1991) The nature of youth tourism: Motivations, characteristics and requirements. Paper presented at the 1991 International Conference on Youth Tourism, New Delhi, World Tourism Organization, Madrid.

Ashley, C. and Haysom, G. (2006) From philanthropy to a different way of doing business: Strategies and challenges in integrating pro-poor approaches into tourism business. *Development Southern Africa* 23, 265–280.

Ateljevic, I. and Doorne, S. (2000) Tourism as an escape: Long-term travellers in New Zealand. *Tourism Analysis* 5, 131–136.

Ateljevic, I. and Doorne, S. (2004) Theoretical encounters: A review of backpacker literature. In G. Richards and J. Wilson (eds) *The Global Nomad: Backpacker Travel in Theory and Practice* (pp. 50–76). Clevedon: Channel View Publications.

Ateljevic, I. and Doorne, S. (2005) Dialectics of authentification: Performing 'exotic otherness' in a backpacker enclave of Dali, China. *Journal of Tourism and Cultural Change* 3 (1), 1–17.

Ateljevic, I. and Hannam, K. (2007) Conclusion: Towards a critical agenda for backpacker tourism. In K. Hannam and I. Ateljevic (eds) *Backpacker Tourism: Concepts and Profiles* (pp. 247–256). Clevedon: Channel View Publications.

Aubert-Gamet, V. and Cova, B. (1999) Servicescapes: From modern non-places to postmodern common places. *Journal of Business Research* 44 (1), 37–45.

Auge, M. (1995) *Non Places: An Introduction to the Non-places of Supermodernity.* London: Verso.

Awesome Adventures Fiji (2008) On WWW at www.awesomefiji.com. Accessed 1.11.08.

Backpackers World Travel On WWW at http://www.backpackersworld.com.au. Accessed 11.08.

Bain, C., Dunford, G., Miller, K., O'Brien, S. and Rawling-Way, C. (eds) (2006) *New Zealand Lonely Planet.* London: Lonely Planet Publications.

Baum, T. and Thompson, K. (2007) Skills and labour markets in transition: A tourism skills inventory of Kyrgystan, Mongolia and Uzbekistan. *Asia Pacific Journal of Human Resources* 45, 235–249.

Bauman, Z. (1998) *Globalization: The Human Consequences.* New York: Columbia University Press.

Bauman, Z. (2001) *Community: Seeking Safety in an Insecure World.* Cambridge: Polity Press.

Baumeister, R.F. 1986, *Identity: Cultural Change and the Struggle for Self.* New York: Oxford University Press.

Bedunah, D.J. and Schmidt, S.M. (2000) Rangelands of Gobi Gurvan Saikhan National Conservation Park, Mongolia. *Rangelands* 22 (4), 18–24.

Bell, C. (2002) The big 'OE': Young New Zealand travellers as secular pilgrims. *Tourist Studies* 2 (2), 143–158.

Bell, C. (2005) The nervous gaze: Backpackers in Africa. In M. Rornero and E. Margolis (eds) *The Blackwell Companion to Social Inequalities* (pp. 424–440). London: Blackwell.

Bell, C. and Lyall, J. (2002) *The Accelerated Sublime: Landscape, Tourism and Identity.* Westport, CT: Praeger Greenwood.

Bell, M. and Ward, G. (2000) Comparing permanent migration with temporary mobility. *Tourism Geographies* 2 (1), 97–107.

Bennett, R.J. (2008) Entering the global margin: Setting the 'Other' scene in independent travel. In P. Burns and M. Novelli (eds) *Tourism and Mobilities: Local-Global Connections* (pp. 133–145). CABI: Trowbridge.

Binder, J. (2004) Travellerscapes. Tourism Research and Transnational Anthropology. Research Group Transnationalism Working Paper Number 6. On WWW at http://www.unifrankfurt.de/fb09/kulturanthro/research/tn/ 1 >. Accessed 12.8.08.

Bisch, J. (1963) *Mongolia: Unknown Land.* New York: E.P. Dutton and Co. Inc.

Bitner, M.J. (1992) Servicescapes: The impact of physical surroundings on customers and employees. *Journal of Marketing* 56 (2), 57–71.

Boden, D. and Molotch, H. (1994) The compulsion to proximity. In R. Friedland and D. Boden (eds) *Nowhere. Space, Time and Modernity* (pp. 257–286). Berkeley, CA: University of California Press.

Bondi, L., Davidson, J. and Smith M. (2007) Introduction: Geography's emotional turn. In J. Davidson, L. Bondi and M. Smith (eds) *Emotional Geographies*. Aldershot: Ashgate.

Boorstin, D. (1964) *The Image: A Guide to Pseudo-Events in America*. New York: Harper.

Bradt, H. (1995) Better to travel cheaply? *The Independent on Sunday Magazine*, 12 February, pp. 49–50.

Breaking Travel News (2006) *The Flashpacker: A new breed of traveller*. 24 March.

Brenner, L. and Fricke, J. (2007) The evolution of backpacker destinations: The case of Zipolite, Mexico. *International Journal of Tourism Research* 9 (3), 217–230.

Britton, S.G. (1982) The political economy of tourism in the third world. *Annals of Tourism Research* 9, 33–358.

Britton, S. (1991) Tourism, capital and place: Towards a critical geography of tourism. *Environment and Planning D: Society and Space* 9, 451–478.

Britton, S. and Clarke, W. (eds) (1987) *Ambiguous Alternative: Tourism in Small Developing Countries*. Suva: University of the South Pacific.

Brown, B. (2007) Working the problems of tourism. *Annals of Tourism Research* 34 (2), 364–383.

Brown, B. and Perry, M. (2002) Of maps and guidebooks: Designing geographical technologies. *Proceedings of Designing Interactive Systems (DIS)*. London: ACM Press.

Brown, S. and Lehto, X. (2005) Travelling with a purpose: Understanding the motives and benefits of volunteer vacationers. *Current Issues in Tourism* 8 (6), 479–496.

Broyard, A. (1982) In praise of contact. *New York Times*, June 27, 1982.

Bruner, E.M. (1991) Transformation of self in tourism. *Annals of Tourism Research* 18, 238–250.

Bruns, A. and Jacobs, J. (2006) *Uses of Blogs*. New York: Peter Lang.

Buckley, R., Ollenburg, C. and Zhong, L. (2008) Cultural landscape in Mongolian tourism. *Annals of Tourism Research* 35 (1), 47–61.

Burns, P. (1999) *An Introduction to Tourism and Anthropology*. London: Routledge.

Bushell, R. (2001) Practice, provision and impact. In N. Douglas, N. Douglas and R. Derrett (eds) *Special Interest Tourism* (pp. 29–55). Queensland, Australia: Wiley and Sons.

Buzzard, J. (1993) *The Beaten Track: European Tourism, Literature and the Ways to "Culture" 1800–1918*. Oxford: Clarendon Press.

Callanan, M. and Thomas, S. (2005) Volunteer tourism. Deconstructing volunteer activities within a dynamic environment. In M. Novelli (ed.) *Niche Tourism. Contemporary Issues, Trends and Cases* (pp. 183–200). Oxford: Butterworth-Heinemann.

Cassim, F. (1993) *Tourism and Development in South Africa*. Cape Town: University of Cape Town Economic Trends Working Paper no. 1.

Castells, M. (1996) *The Rise of the Network Society, The Information Age: Economy, Society and Culture* (Vol. I). Cambridge, MA; Oxford: Blackwell.

Cave, J., Thyne, M. and Ryan, C. (2007) Perceptions of backpacker accommodation facilities: A comparative study of Scotland and New Zealand. In

K. Hannam and I. Ateljevic (eds) *Backpacker Tourism: Concepts and Profiles* (pp. 215–246). Clevedon: Channel View Publications.

Cederholm, E. A. (2004) The use of photo-elicitation in tourism research – Framing the backpacker experience. *Scandinavian Journal of Hospitality and Tourism* 4 (3), 225–241.

Cheong, S. and Miller, M. (2000) Power and tourism: A Foucauldian observation. *Annals of Tourism Research* 27 (2), 371–390.

Clarke, N. (2004a) Mobility, fixity, agency: Australia's working holiday programme population. *Space and Place* 10, 411–420.

Clarke, N. (2004b) Free independent travellers? British working holiday makers in Australia. *Transactions of the Institute of British Geographers* 29 (4), 499–509.

Clarke, N. (2005) Detailing transnational lives of the middle: British working holiday makers in Australia. *Journal of Ethnic and Migration Studies* 31 (2), 307–322.

Clift, S. and Forrest, S. (1999a) Gay men and tourism: Destinations and holiday motivations. *Tourism Management* 20 (5), 615–625.

Clift, S. and Forrest, S. (1999b) Factors associated with gay men's sexual behaviours and risk on holiday. *AIDS Care* 11 (3), 281–295.

Clift, S., Lugano, M. and Callister, C. (eds) (2002) *Gay Tourism: Culture, Identity and Sex*. New York: Continuum.

Coasting Africa (2008) *Coast to Coast Guide 2008*. On WWW at www.coastingafrica.com. Accessed 15.4.08.

Cochrane, J. (2005) The backpacker plus: Overlooked and underrated. Paper read at the ATLAS Expert Meeting, Bangkok, September.

Cohen, E. (1972) Toward a sociology of international tourism. *Social Research* 39 (1), 164.

Cohen, E. (1973) Nomads from affluence: Notes on the phenomenon of drifter-tourism. *International Journal of Comparative Sociology* 14 (1), 89–103.

Cohen, E. (1974) Who is a tourist? A conceptual classification. *Sociological Review* 22, 527–555.

Cohen, E. (1982a) Marginal paradises. Bungalow tourism on the islands of Southern Thailand. *Annals of Tourism Research* 9, 189–228.

Cohen, E. (1982b) Thai girls and farang men: The edge of ambiguity. *Annals of Tourism Research* 9 (3), 403–428.

Cohen, E. (2003) Backpacking: Diversity and change. *Journal of Tourism & Cultural Change* 1 (2), 95–110.

Cohen, E. (2004a) Backpacking: Diversity and change. In G. Richards and J. Wilson (eds) *The Global Nomad: Backpacker Travel in Theory and Practice* (pp. 43–59). Clevedon: Channel View Publications.

Cohen, E. (2004b) *Contemporary tourism: Diversity and change*. Boston, MA: Elsevier.

Coles, T., Duval, T. and Hall, C.M. (2005) Tourism, mobility and global communities: New approaches to theorising tourism and tourist spaces. In W. Theobald (ed.) *Global Tourism* (3rd edn) (pp. 463–481). London: Elsevier.

Coles, T. and Hall, C.M. (2006) The geography of tourism is dead. Long live geographies of tourism and mobility. *Current Issues in Tourism* 9 (4–5), 289–319.

Coles, T. and Timothy, D.J. (2004) *Tourism, Diasporas and Space: Travels to Promised Lands*. London: Routledge.

Commonwealth of Australia (2003) *Tourism White Paper*. Canberra: AGPS.

Cornelissen, S. (2005) *The Global Tourism System: Governance, Development and Lessons from South Africa.* Aldershot: Ashgate.

Cooper, M., O'Mahoney, K. and Erfurt, P. (2004) Backpackers: Nomads join the mainstream? An analysis of backpacker employment on the Harvest Trail Circuit in Australia. In G. Richards and J. Wilson (eds) *The Global Nomad: Backpacker Travel in Theory and Practice* (pp. 180–195). Clevedon: Channel View Publications.

Crawshaw, C. and Urry, J. (1997) Tourism and the photographic eye. In C. Rojek and J. Urry (eds) *Touring Cultures, Transformations of Travel and Theory* (pp. 45–67). London: Routledge.

Crang, M. (2003) Qualitative methods: Touchy, feely, look-see. *Progress in Human Geography* 29, 225–233.

Crouch, D. (1999) Introduction. In D. Crouch (ed.) *Leisure/Tourism Geographies: Practices and Geographical Knowledge* (pp. 1–14). London. Routledge.

Crouch, D. and Desforges, L. (2003) The sensuous in the tourist encounter: Introduction: The power of the body in tourist studies. *Tourist Studies* 3 (1), 5–22.

D'andrea, A. (2007) *Global Nomads Techno and New Age as Transnational Counter-cultures in Ibiza (Spain) and Goa (India).* London: Routledge.

Dann, G. (1994) Travel by train: Keeping nostalgia on track. In A.V. Seaton (ed.) *Tourism: The State of the Art* (pp. 775–783). Chichester: Wiley.

Dann, G. (1999) Writing out the tourist in space and time. *Annals of Tourism Research* 26 (1), 159–187.

Davidson, K. (2005) Alternative India: Transgressive spaces. In A. Jaworski and A. Pritchard (eds) *Discourse, Communication and Tourism* (pp. 28–52). Clevedon: Channel View Publications.

DeCerteau (1988) *The Practice of Everyday Life.* Berkeley, CA: The University of Berkeley Press.

DeFrancis, J. (1993) *In the Footsteps of Ghengis Khan.* Honolulu, HI: University of Hawaii Press.

Desforges, L. (1998) 'Checking out the planet': Global representations/local identities and youth travel. In T. Skelton and G. Valentine (eds) *Cool Places: Geographies of Youth Cultures* (pp. 175–192). London: Routledge.

Desforges, L. (2000) Traveling the world: Identity and travel biography. *Annals of Tourism Research* 27 (4), 926–945.

Delanty, G. (2003) *Community.* London and New York: Routledge.

Department of Environmental Affairs and Tourism (DEAT) (2008) *A Human Resource Development Strategy for the Tourism Sector.* Pretoria: DEAT.

Department of Trade and Industry (DTI) (2005a) *A Growing Economy that Benefits All: Accelerated & Shared Growth Initiative for South Africa (ASGI-SA): Discussion Document.* Pretoria: DTI.

Department of Immigration and Multicultural Affairs (DIMA) (2008) *Fact Sheet: Working Holiday Maker Programme* [online]. On WWW at http://www.immi.-gov.au/media/fact-sheets/49whm.htm. Accessed 3.09.

Discovery Coast Tourism & Commerce (2006) *Captain Cook – Town of 1770 – Agnes Water.* On WWW at http://www.townof1770-agneswater.com.au/history.htm.

Donaldson, J.A. (2007) Tourism development and poverty reduction in Guizhou and Yunnan. *China Quarterly* 190, 333–351.

Doorne, S. and Ateljevic, I. (2005) Tourism performance as metaphor: Enacting backpacker travel in the Fiji. In A. Jaworski and A. Pritchard (eds) *Discourse, Communication and Tourism* (pp. 173–198). Clevedon: Channel View Publications.

DTI (2005b) *Tourism Sector Development Strategy*. Pretoria: Trade and Investment South Africa for DTI.

DTI (2007a) *Industrial Policy Action Plan (IPAP)*. Pretoria: DTI.

DTI (2007b) *Backpacking and Youth Travel in South Africa*. Pretoria: DTI.

DTI (2008) Tourism news, *DTI Tourism e-news* Quarter 1-March. On WWW at http://www.thedti.gov.za/publications/tourismnewsmarch08.htm. Accessed 31.3.08.

Dunn, G. (2004) *Swagman #14 – Noosa and the Development of a Backpacker Town*. On WWW at http://www.bootsnall.com/travelogues/gdunn/14.shtml. Accessed 6.12.08.

Duncan, M. (2004) Autoethnography: Critical appreciation of an emerging art. *International Journal of Qualitative Methods* 3 (4), 1–8.

Edensor, T. (1998) *Tourists at the Taj: Performance and Meaning at a Symbolic Site*. London: Routledge.

Edensor, T. (2000) Staging tourism: Tourists as performers. *Annals of Tourism Research* 27 (2), 322–344.

Edensor, T. (2001) Performing tourism, staging tourism: (Re)producing tourist space and practice. *Tourist Studies* 1 (1), 59–81.

Edensor, T. (2004a) Reconstituting the Taj Mahal: Tourist flows and glocalization. In M. Sheller and J. Urry (eds) *Tourism Mobilities* (pp. 103–115). New York: Routledge.

Edensor, T. (2004b) Automobility and national identity. *Theory, Culture & Society* 21 (4–5), 101–120.

Edensor, T. (2007) Mundane mobilities, performances and spaces of tourism. *Social & Cultural Geography* 8 (2), 199–215.

Ek, R. and Hultman, J. (2008) Sticky landscapes and smooth experiences: The biopower of tourism mobilities in the Öresund Region. *Mobilities* 3 (2), 223–242.

Elsrud, T. (1998) Time creation in travelling: The taking and making of time among women backpackers. *Time and Society* 7 (2), 309–334.

Elsrud, T. (2001) Risk creation in travelling: Backpacker adventure narration. *Annals of Tourism Research* 28 (3), 597–617.

Ferrario, F.F. (1978) *An Evaluation of the Tourist Resources of South Africa*. Cape Town: Publication 1, Department of Geography, University of Cape Town.

*Flashpackingwife.com.* (2009) *Flashpackers Return Home with Flashbaby!* On WWW at http://www.flashpackingwife.com/flashpackers-return-home-with-flash-baby/. Accessed 19.7.09.

Florida, R. (2005) *Cities and the Creative Class*. London: Routledge.

Footprint. (2004) *Australia*. Footprint: Bath.

Ford, M.J. (2007) *Agnes Water Revealed!* Online source. Accessed 24.8.8.

Foucault, M. (1975) *Discipline and Punish: The Birth of the Prison*. London, Penguin.

Foucault, M. (1980) *The History of Sexuality*, Vol 1. An Introduction. New York: Vintage.

Foucault, M. (1988) Technologies of the self. In L.H. Martin, H. Gutman and P.H. Hutton (eds) *Technologies of the Self* (pp. 16–49). Amherst, MA: University of Massachusetts Press.

Franklin, A. (2003) *Tourism: An Introduction*. London: Sage.

Franklin, A. and Crang, M. (2001) The trouble with tourism and travel theory? Editorial. *Tourist Studies* 1 (1), 5–22.

Fussell, P. (1980) *Abroad: British Literary Traveling Between the Wars.* New York: Oxford University Press.

Gogia, N. (2006) Unpacking corporeal mobilities: The global voyages of labour and leisure. *Environment and Planning* 38, 359–375.

Garland, A. (1996) *The Beach.* Harmondsworth: Penguin.

Galani-Moutafi, V. (2000) The Self and the Other: Traveler, ethnographer, tourist. *Annals of Tourism Research* 27 (1), 203–224.

Gibson, H. and Yiannakis, A. (2002) Tourist roles: Needs and the lifecourse. *Annals of Tourism Research* 29, 358–383.

Giddens, A. (1991) *Modernity and Self-Identity: Self and Society in the Late Modern Age.* Cambridge: Polity Press.

Gladstone Area Promotion and Development Limited (2005) *The Gladstone Region Project Status Report – Update June 2005.* On WWW at *www.aph.gov.au/HOUSE/ committee/trs/networks/subs/sub084b.pdf.* Gladstone.

Gladstone Area Promotion and Development Limited (2006) *Gladstone Regional Overview – September Quarter 2006.* On WWW at http://svc105.wic466d.serverweb.com/images/gladstone/gladstone%20region%20overview%20-%20sept%202006.pdf.

Goffman, E. (1963) *Behavior in Public Places: Notes on the Social Organization of Gatherings.* London: Free Press.

Goffman, E. (1977) *Relations in Public.* London: Penguin.

Graburn, N.H.H. (1977) Tourism: The sacred journey. In V.L. Smith (ed.) *Hosts and Guests: The Anthropology of Tourism* (pp. 17–31). Philadelphia, PA: University of Pennsylvania Press.

Graburn, N.H.H. (1983) The anthropology of tourism. *Annals of Tourism Research* 10 (1), 9–33.

Graburn, N.H.H. (1989) Tourism: The sacred journey. In V. Smith (ed.) *Hosts and Guests: The Anthropology of Tourism* (2nd edn; pp. 21–36). Philadelphia, PA: University of Pennsylvania Press.

Graham, S. (2004) Introduction: From dreams of transcendence to the remediation of urban life. In S. Graham (ed.) *The Cybercities Reader* (pp. 1–30). London: Routledge.

Gunn, C. (1989) *Vacationscape: Designing Tourist Regions* (2nd edn). New York: Van Nostrand Reinhold.

Hall, C.M. (2005a) Reconsidering the geography of tourism and contemporary mobility. *Geographical Research* 43 (2), 125–139.

Hall, C.M. (2005b) *Tourism: Rethinking the Social Science of Mobility.* London: Prentice Hall.

Hall, C.M. and Page, S.J. (2009) Progress in tourism management: From the geography of tourism to geographies of tourism – a review. *Tourism Management* 30, 3–16.

Hall, C.M. and O'Sullivan, V. (1996) Tourism, political instability & social unrest. In A. Pizam and Y. Mansfeld (eds) *Tourism, Crime and International Security Issues* (pp. 105–21). London: Wiley.

Hall, D. (2001) Tourism and development in communist and post-communist societies. In D. Harrison (ed.) *Tourism and the Less Developed World: Issues and Case Studies* (pp. 91–107). Wallingford: CABI.

Hampton, M.P. (1998) Backpacker tourism and economic development. *Annals of Tourism Research* 25 (3), 639–660.

Hampton, M.P. (2003) Entry points for local tourism in developing countries: Evidence from Yogyakarta, Indonesia. *Geografiska Annaler B: Human Geography* 85 (2), 85–101.

Hampton, M.P. (2005) Heritage, local communities and economic development. *Annals of Tourism Research* 32 (3), 735–759.

Hampton, M. and Hampton, J. (2009) Is the Beach Party over? Tourism and the environment in a small island: A case study of Gili Trawangan, Lombok, Indonesia. In M. Hitchcock, V.T. King and M. Parnwell (eds) *Tourism in South-East Asia: Challenges and New Directions* (pp. 286–308). Copenhagen: NIAS Press.

Hampton, M.P. and Hamzah, A. (2008) The changing geographies of backpacker tourism in south-east Asia. Paper read at the ATLAS expert conference on backpacker tourism, Shimla, India; March.

Hamzah, A. (1995) The changing tourist motivation and its implications on the sustainability of small-scale tourism development in Malaysia. Paper read at World Conference on Sustainable Tourism, Lanzarote, Spain, 24–29 April.

Hamzah, A. and Hampton, M.P. (2007) Reaching tipping point? Tourism development, governance and the environment in small islands: Lessons from Perhentian Kecil, Malaysia. Paper read at IGU Commission on Islands 'Island Geographies' Conference, Taipei, Taiwan 29 October to 3 November.

Hannam, K. (2006) Tourism and development III: Performances, performativities and mobilities. *Progress in Development Studies* 6 (3), 243–249.

Hannam, K. (2008) Tourist geographies, tourist studies and the turn towards mobilities. *Geography Compass* 2 (1), 127–139.

Hannam, K. and Ateljevic, I. (2007a) *Backpacker Tourism: Concepts and Profiles*. Clevedon: Channel View Publications.

Hannam, K. and Ateljevic, I. (2007b) Introduction: Conceptualising and profiling backpacker tourism. In K. Hannam and I. Ateljevic (eds) *Backpacker Tourism: Concepts and Profiles* (pp. 1–6). Clevedon: Channel View Publications.

Hannam, K., Sheller, M. and Urry, J. (2006) Editorial: Mobilities, immobilities and moorings. *Mobilities* 1, 1–22.

Hannigan, J. (2007) A neo-Bohemian rhapsody: Cultural vibrancy and controlled edge as urban development tools in the 'new creative economy'. In T.A. Gibson and M. Lowes (eds) *Urban Communication: Production, Text, Context* (pp. 61–81). New York: Rowan and Littlefield.

Harding, G. and Webster, E. (2002) *The Working Holiday Maker Scheme and The Australian Labour Market*. On WWW at http://www.immi.gov.au/media/publications/research/whm/. Accessed 3.09.

Haviv, A.S. (2005) Next year in Kathmandu: Israeli backpackers and the formation of a new Israeli identity In C. Noy and E. Cohen (eds) *Israeli Backpackers and their Society: A View from Afar* (pp. 45–88). Albany, NY: State University of New York.

Heaney, L. (2003) *Bureau of Tourism Research Niche Market Report Number 2: Backpackers in Australia 2002*. Canberra: Bureau of Tourism Research.

Henning, H. (1946) *Mongolian Journey*. London: Routledge and Kegan Paul.

Hesse, H. (1964) *The Journey to the East*. London: Peter Owen.

Hetherington, K. (1988) *Expressions of Identity*. London: Sage.

Hitchcock, M., King, V.T and Parnwell, M. (eds) *Tourism in South-East Asia: Challenges and New Directions.* Copenhagen: NIAS Press.

Holt, N.L. (2003) Representation, legitimation, and autoethnography: An autoethnographic writing story. *International Journal of Qualitative Methods* 2 (1), 1–22.

Hostelworld.com (2009) Luxury hostels. On WWW at http://www.hostelworld.com/travel-features/80/luxury-hostels/?source = googleadwordshostels&gcl id = CNb4uvDY4psCFYYU4wodLW2__g. Accessed 19.7.09.

Hottola, P. (2004) Culture confusion – intercultural adaptation in tourism. *Annals of Tourism Research* 31 (2), 447–466.

Hottola, P. (2005) The metaspatialities of control management in tourism: Backpacking in India. *Tourism Geographies* 7 (1), 1–22.

Howard, R.W. (2007) Five backpacker tourist enclaves. *International Journal of Tourism Research* 9 (2), 73–86.

Howe, A.C. (2001) Queer pilgrimage: The San Francisco homeland and identity tourism. *Cultural Anthropology* 16 (1), 35–61.

Hudson, R. (2005) *Economic Geographies: Circuits, Flows and Spaces.* London: Sage.

Hughes, H. (1997) Holidays and homosexual identity. *Tourism Management* 18 (1), 3–7.

Hughes, H. (2006) *Pink Tourism: Holidays of Gay Men and Lesbians.* Wallingford: CABI.

Hutnyk, J. (1996) *The Rumour of Calcutta. Tourism. Charity and the Poverty of Representation.* London: Zed Books.

Hutnyk, J. (2007) The banana pancake trail. Trinketization. On WWW at http://hutnyk.blogspot.com/2007/07/banana-pancake-trail.html. Blog posted on 8 July.

Hutnyk, J. (2008) Writing diary. Trinketization: Anthropology. On WWW at http://hutnyk.blogspot.com/search/label/anthropology. Blog posted on 2 March.

Huxley, L. (2007) Western backpackers and the global experience: An exploration of young people's interaction with local cultures. *Tourism, Culture & Communication* 5, 37–44.

Hyde, K. (2008) Information processing and touring planning theory. *Annals of Tourism Research* 35 (3), 712–731.

Ianchovichina, S. and Gooptu (2007) Growth diagnostics for a resource-rich transition economy: The case of Mongolia. World Bank Policy Research Working Paper, No. 4396.

Inda, J.X. and Rosaldo, R. (eds) (2008) *The Anthropology of Globalisation. A reader* (2nd edn). Oxford: Blackwell.

Ipalawatte, C. (2004) *Backpackers in Australia 2003: Niche Market Report Number 4.* Canberra: Tourism Research Australia.

Iyer, P. (1988) *Video Night in Kathmandu and Other Reports from the Not-so-Far East.* London: Black Swan.

Jacobsen, J.K.S. (2000) Anti-tourist attitudes: Mediterranean charter tourism. *Annals of Tourism Research* 27 (2), 284–300.

Jacobsen, J.K.S. (2001) Nomadic tourism and fleeting place encounters: Exploring different aspects of sightseeing. *Scandinavian Journal of Hospitality and Tourism* 1 (2), 99–112.

Jacobsen, J.K.S. (2004) Roaming romantics: Soltitude-seeking and self-centredness in scenic sightseeing. *Scandinavian Journal of Hospitality and Tourism* 4 (1), 5–23.

Jamal, T. and Hill, S. (2002) The home and the world; (post)touristic spaces of (in)authenticity? In G. Dann (ed.) *The Tourist as a Metaphor of the Social World* (pp. 77–107). Wallingford: CABI.

Jamieson, K. (1996) 'Been there–done that' Identity and the Overseas Experiences of Young Pakeha New Zealanders, Masters thesis. Massey University.

Jansson, A. (2007) A sense of tourism: New media and the dialectic of encapsulation/decapsulation. *Tourist Studies* 7 (1), 5–24.

Jarvis, J. (2004) Yellow bible tourism: Backpackers in Southeast Asia. In B. West (ed.) *Down the Road: Exploring Backpackers and Independent Travel* (pp. 153–167). Australia: API Network.

Jeffrey, S. (2003) Do women do it too? *Leisure Studies* 22, 223–238.

Jenkins, R. (1996) *Social Identity.* London: Routledge.

Johnston, L. (2001) (Other) bodies and tourism studies. *Annals of Tourism Research* 28 (1), 180.

Jordan, F. and Gibson, H. (2005) 'We're not stupid...but we'll not stay home either': Experiences of solo women travellers. *Tourism Review International* 9, 195–211.

Judd, D. (2003) Visitors and the spatial ecology of the city. In L.K. Hoffman, S.S. Fainstein and D.R. Judd (eds) *Cities and Visitors: Regulating People, Markets, and City Space* (pp. 23–38). New York: Blackwell Publishing.

Kanta, V. (2002) 'Certain places have different energy': Spatial transformations in Eresos, Lesvos. GLQ, *A Journal of Gay and Lesbian Studies* 8 (1–2), 35–56.

Keesing, R.M. (1974) Theories of culture. *Annual Review of Anthropology* 3, 73–97.

Kerouac, J. (1957) *On the Road.* New York: Penguin.

Kinnaird, V. and Hall, D. (1994) *Tourism: A Gender Analysis.* Chichester: Wiley.

Larsen, J. (2001) Tourism mobilities and the travel glance: Experiences of being on the move. *Scandinavian Journal of Hospitality and Tourism* 1 (2), 80–98.

Larsen, J., Urry, J. and Axhausen (2007) Networks and tourism: Mobile social life. *Annals of Tourism Research* 34 (1), 244–262.

Lash, S. and Urry, J. (1994) *Economies of Signs and Space.* London. Sage.

Lassen, C. (2006) Aeromobility and work. *Environment and Planning A* 38 (2), 301–312.

Lawless, J. (2000) *Wild East. Travels in the New Mongolia.* Chichester: Summersdale.

Lee, Tze I. and Ghazali, M. (2007) Uncovering the international backpacker to Malaysia. In K. Hannam and I. Ateljevic (eds) *Backpacker Tourism: Concepts and Profiles* (pp. 128–143). Clevedon: Channel View Publications.

Leiper, N. (1990) Tourist attraction systems. *Annals of Tourism Research* 17 (3), 367–384.

Lepp, A. and Gibson, H. (2003) Tourist roles, perceived risk and international tourism. *Annals of Tourism Research* 30 (3), 606–624.

Let's Connect (2007) *1770 Festival.* On WWW at http://www.letsconnect.com.au/1770/1770festival-pp2.htm.

Lett, J.W. (1983) Ludic and liminoid aspects of charter yacht tourism in the Caribbean. *Annals of Tourism Research* 10 (1), 35–56.

Levett, R. and McNally, R. (2003) A strategic environmental assessment of Fiji's development plan. World Wide Fund for Nature, South Pacific, Suva, Fiji.

Lloyd, K. (2003) Contesting control in transitional Vietnam: The development and regulation of traveller cafes in Hanoi and Ho Chi Minh City. *Tourism Geographies* 5 (3), 350–366.

Lofgren, O. (2002) *On Holiday*. Berkley, CA: University of California Press.

Loker, L. (1993) *The Backpacker Phenomenon II: More Answers to Further Questions*. Townsville: James Cook University of North Queensland.

Loker-Murphy, L. and Pearce, P. (1995) Young budget travelers: Backpackers in Australia. *Annals of Tourism Research* 22 (4), 819–843.

Lonely Planet (2007) *Australia*. Footscray.

Longhurst, R., Ho, E. and Johnston, L (2008) Using 'the body' as an 'instrument of research': kimch'i and pavlova. *Area* 40 (2), 208–217.

Lowitt, S. (2006) Sector studies and employment scenarios: A view of South Africa's tourism sector – preliminary findings. Report prepared for the Human Sciences Research Council, Pretoria.

Lupton, D. (1996) *Food, the Body and the Self*. London: Sage.

Luvsandsvaajav, O. (2005) Mongolia's image as a tourist destination – perception of foreign tourists. Paper presented at The International Conference on Destination Marketing and Branding for Regional Tourism Development, Macao S.A.R., China, 8–10 December.

Lyons, G. and Urry, J. (2005) Travel time use in the information age. *Transportation Research Part A* 39, 257–276.

Macbeth, J. (2000) Utopian tourists – cruising is not just about sailing. *Current Issues in Tourism* 3 (1), 20–34.

Macbeth, J. and Westerhausen, K. (2001) The development of backpacker tourism in Western Australia. On WWW at http://wwwsoc.murdoch.edu.au/tourism/backpacker.htm.

Macbeth, J. and Westerhausen, K. (2001) *The Development of Backpacker Tourism in Western Australia*. On WWW at http://tourism.murdoch.edu.au/backpacker.htm. Accessed 16.10.01.

Macbeth, J. and Westerhausen, K. (2003) Backpackers and empowered local communities: Natural allies in the struggle for sustainability and local control? *Tourism Geographies* 5 (1), 71–86.

MacCannell, D. (1999) *The Tourist: A New Theory of the Leisure Class* (3rd edn). Berkeley, CA: University of California Press.

MacLean, R. (2006) *Magic Bus: On the Hippie Trail from Istanbul to India*. London: Viking.

MacKay, K. and Fesenmaier, D. (1997) Pictorial element of destination in image formation. *Annals of Tourism Research* 24, 537–565.

Maffesoli, M. (1996) *The Time of the Tribes*. London: Sage.

Malbon, B. (1998) The club: Clubbing: Consumption, identity and the spatial practices of every-night life. In T. Skelton and G. Valentine (eds) *Cool Places: Geographies of Youth Cultures* (pp. 266–286). London: Routledge.

Mascheroni, G. (2007) 'Global nomads' network and mobile sociality: Exploring new media uses on the move. *Information, Communication, and Society* 10 (4), 527–546.

Maoz, D. (2004) The conquerors and the settlers: Two groups of young Israeli backpackers in India. In G. Richards and J. Wilson (eds) *The Global Nomad: Backpacker Travel in Theory and Practice* (pp. 109–122). Clevedon: Channel View Publications.

Maoz, D. (2006) The mutual gaze. *Annals of Tourism Research* 33 (1), 221–239.

Maoz, D. (2007) Backpackers' motivations: The role of culture and nationality. *Annals of Tourism Research* 34 (1), 122–140.

Maoz, D. (2007) The backpacking journey of Israeli women in mid-life. In K. Hannam and I. Ateljevic (eds) *Backpacker Tourism: Concepts and Profiles* (pp. 188–198). Clevedon: Channel View Publications.

Maxwell, J. (2008) Packed with experience. *Australian Financial Review Magazine* 17, 42–44.

McIntosh, A. and Zahra, A. (2007) A cultural encounter through volunteer tourism: Towards the ideals of sustainable tourism? *Journal of Sustainable Tourism* 15 (5), 541–556.

Mehmetoglu, M., Dann, G.M.S. and Larsen, S. (2001) Solitary travellers in the Norwegian Lofoten Islands: Why do people travel on their own? *Scandinavian Journal of Hospitality and Tourism* 1 (1), 19–37.

Metcalf, B. (1995) *From Utopian Dreaming to Communal Reality: Cooperative Lifestyles in Australia*. Sydney: University of New South Wales Press.

Meyer, L. (2008) Event report – The World Youth and Student Travel Conference 2007 (WYSTC). On WWW at www.backpackingsouthafrica.co.za/index.php?option = com_content&task = view. Accessed 15.10.08.

Michener, J.A. (1971) *The Drifters*. London: Random House.

Miles, P. (2004) Best of Both Worlds. *The Guardian*, 12 June.

Ministry of Information, Communications and Media Relations (2004) Fiji Today 2004–2005. On WWW at http://www.beta.fiji.gov.fj/uploads/FijiToday2004-5.pdf.

Ministry of Tourism (2006) International Visitor Survey. Annual Report, Fiji.

Ministry of Tourism (2007) *Study on the Contribution and Potential of Backpacker Tourism in Malaysia*. Kuala Lumpur: Ministry of Tourism Malaysia.

Mograbi, J. (2007) The role of Gauteng in the South African backpacker economy. Unpublished MSc dissertation, University of the Witwatersrand.

Mokhtarian, P.L. (2005) Travel as a desired end, not just a means. *Transportation Research Part A: Policy and Practice* 39 (2–3), 93–96.

Molz, J.G. (2005) Getting a 'flexible eye': Round-the world travel and scales of cosmopolitan citizenship. *Citizenship Studies* 9 (5), 518.

Molz, J.G. (2006) Watch us wander: Mobile surveillance and the surveillance of mobility. *Environment and Planning A* 38 (2), 377–393.

Morris, M. (2006) *Identity Anecdotes: Translation and Media Culture*. London: Sage.

Moshin, A. and Ryan, C. (2003) Backpackers in the Northern Territory of Australia – motives, behaviours and satisfactions. *International Journal of Tourism Research* 5, 113–131.

More, T. (1516) 1910. *Utopia*. London: Everyman's Library.

Moore, J. (2004) *Visions of Culture: An Introduction to Anthropological Theories and Theorists* (2nd edn). Lanham, MD: Rowman Altamira.

Mowforth, M. and Munt, I. (2003) *Tourism and Sustainability: Development and New Tourism in the Third World* (2nd edn). London: Routledge.

Munro, P. (2008) Desperately seeking a tourist trap. *Sun Herald* 29 June, p. 17.

Murphy, L.E. (2001) Exploring social interactions of backpackers. *Annals of Tourism Research* 26, 50–67.

Muzaini, H. (2006) Backpacking Southeast Asia: Strategies of 'looking local'. *Annals of Tourism Research* 33 (1), 144–161.

Myers, L. and Hannam, K. (2007) Women as backpacker tourists: A feminist analysis of destination choice and social identifies form the UK. In K. Hannam and I. Ateljevic (eds) *Backpacker Tourism: Concepts and Profiles* (pp. 174–187). Clevedon: Channel View Publications.

Myslik, W.D. (1996) Re-negotiating the social/sexual identities of place: Communities as safe havens or sites of resistance. In N. Duncan (ed.) *Bodyspace Destabilizing Geographies of Gender and Sexuality* (pp. 156–169). New York: Routledge.

Nash, D. (2001) On travelers, ethnographers and tourists. *Annals of Tourism Research* 28 (2), 493–496.

Newlands, K. (2004) Setting out on the road less travelled: A study of backpacker travel in New Zealand. In G. Richards and J. Wilson (eds) *The Global Nomad: Backpacker Travel in Theory and Practice* (pp. 217–236). Clevedon: Channel View Publications.

Newlands, K.J. (2006) The modern nomad in New Zealand: A study of the effects of the working holiday schemes on free independent travellers and their host communities. Unpublished Master of Business (Tourism) thesis, Auckland University of Technology.

Niggel, C. and Benson, A. (2007) Exploring the motivations of backpackers: The case of South Africa. In K. Hannam and I. Ateljevic (eds) *Backpacker Tourism: Concepts and Profiles* (pp. 144–156). Clevedon: Channel View Publications.

Nimmo, K. (2001) Willing workers on organic farms: A case study. Unpublished MA thesis, Victoria University of Wellington.

Nordstrom, J. (2004) Estimating and predicting international tourism demand in Sweden. *Scandinavian Journal of Hospitality and Tourism* 4 (1), 59–76.

Normark, D. (2006) Tending to mobility: Intensities of staying at the petrol station. *Environment & Planning A* 38 (2), 241–252.

Noy, C. (2004) This trip really changed me: Backpackers' narratives of self-change. *Annals of Tourism Research* 31 (1), 78–102.

Noy, C. and Cohen, E. (2005) Introduction: Backpacking as a rite of passage in Israel. In C. Noy and E. Cohen (eds) *Israeli Backpackers and Their Society: A View from Afar* (pp. 7–49). New York: University of New York Press.

Obenour, W., Patterson, M., Pedersen P. and Pearson, L. (2006) Conceptualization of a meaning-based research approach for tourism service experiences. *Tourism Management* 27 (1), 34–41.

Obrador-Pons, P. (2003) Being-on-holiday: Tourist dwelling, bodies and place. *Tourist Studies* 3 (1), 47–66.

Oppermann, M. (1999) Sex tourism. *Annals of Tourism Research* 26 (2), 251–266.

O'Regan, M. (2008) Hypermobility in backpacker lifestyles: The emergence of the Internet care. In P. Burns and M. Novelli (eds) *Tourism and Mobilities: Local-Global Connections* (pp. 109–132). Trowbridge: CABI.

O'Reilly, C. (2005) Tourist or traveller? Narrating backpacker identity. In A. Jaworski and A. Pritchard (eds) *Discourse, Communication and Tourism* (pp. 150–169). Clevedon: Channel View Publications.

O'Reilly, C. (2006) From drifter to gap year tourist: Mainstreaming backpacker travel. *Annals of Tourism Research* 33 (4), 998–1017.

Osborn, R. (2006) *Around Agnes Water/1770*. On WWW at http://www.realestate.com.au/doc/locality/au-qld-agneswater1770.htm.

Oz Experience (2008) *Agnes Water Town of 1770 Travel and Tour Information*. On WWW at http://www.ozexperience.com/agnes-waters-town-of-177.

Oz Experience (2009) *Agnes Water Town of 1770 Travel and Tour Information*. On WWW at http://www.ozexperience.com/agnes-waters-town-of-177.

Paris, C. (2008) *The Backpacker Market: Targeting a Mobile Population Through Online Communities*. Berlin: VDM-Verlag.

Paris, C. and Teye, V. (2010) Understanding backpacker motivations: A travel career approach. *Journal of Hospitality Marketing & Management* (forthcoming) 19 (3).

Paur, J.K. (2002) Queer tourism. Geographies of globalisation. *A Journal of Gay and Lesbian Studies* 8 (1–2), 1–6.

Pearce, P.L. (1990) *The Backpacker Phenomenon: Preliminary Answers to Basic Questions*. Townsville: James Cook University.

Pearce, P.L. (2006) Backpacking and backpackers: A fresh look. *Tourism Recreation Research* 31 (3), 5–10.

Pearce, P. and Foster, F. (2007) A 'University of Travel': Backpacker learning. *Tourism Management* 28, 1285–1298.

Peel, V. and Steen, A. (2007) Victims, hooligans and cash-cows: Media representations of the international backpacker in Australia. *Tourism Management* 28 (4), 1057–1067.

Pooley, C., Turnbull, J. and Adams, M. (2005) *A Mobile Century?: Changes in Everyday Mobility in Britain in the Twentieth Century*. Aldershot: Ashgate.

Poria, Y. (2006) Assessing gay men and lesbian women's hotel experiences: An exploratory study of sexual orientation in the travel industry. *Journal of Travel Research* 44, 327–334.

Poria, Y., Butler, R. and Airey, D. (2003) The core of heritage tourism. *Annals of Tourism Research* 3 (1), 238–254.

Pratt, M.L. (1992) *Imperial Eyes: Travel Writing and Transculturation*. London: Routledge.

Prentice, R. (2004) Tourist familiarity and imagery. *Annals of Tourism Research* 31 (4), 923–945.

Prideaux, B. and Coghlan, A. (2006a) Wildlife tourism in tropical North Queensland: An overview of visitor preferences for wildlife experiences. Tourism Monograph Series No 3, James Cook University Cairns.

Prideaux, B. and Coghlan, A. (2006b) *Backpacking Shopping in the Tropics: An Overview of the Shopping Behaviour of Backpackers in Cairns*. Cairns: James Cook University Research Report 2.

Prideaux, B. and Shiga, H. (2007) Japanese backpacking: The emergence of a new market sector – a Queensland case study. *Tourism Review International* 11, 45–56.

Pritchard, A. and Morgan, J. (1999) The gay consumer: A meaningful market segment? *Journal of Targeting, Measurement and Analysis of Marketing* 6 (1), 9–20.

Pritchard, A., Morgan, J., Sedgley, D., Khan, E. and Jenkins, A. (2000) Sexuality and holiday choices: Conversations with gay and lesbian tourists. *Leisure Studies* 19 (4), 267–282.

Pritchard, A., Morgan, J. and Sedgley (2007) In Search of Lesbian Space? The Experience of Manchester's Gay Village. In A. Pritchard, N. Morgan, I. Ateljevic and C. Harris (eds) *Tourism and Gender: Embodiment, Sensuality and Experience*. Wallingford: CABI.

Pursall, R. (2005) From backpacker to flashpacker. Conference Paper ATLAS Sig meeting Backpackers Research Group, The Global Nomad – an Expert

Meeting on Backpacker Tourism, Kasetsart University, Bangkok, Thailand, 1–3 September.

Queensland Government (2007) *Queensland population update including regional population trends – no. 8, March 2007*. On WWW at http://www.dip.qld.gov.au/docs/planning/planning/information_and_forecasting/QPU-mar-2007/full-report.pdf.

Reisinger and Mavondo (2005) Travel anxiety and intentions to travel internationally: Implications of travel risk perception. *Journal of Travel Research* 43 (3), 212–225.

Richards, G. (2007) *New Horizons in Independent Youth and Student Travel, a Report for the International Student Travel Confederation (ISTC) and the Association of Tourism and Leisure Education (ATLAS)*. Amsterdam: ISTC.

Richards, G. (2007) *New Horizons 11 – The Young Independent Traveller*. The World Youth Student and Educational (WYSE) Travel Confederation: Amsterdam. On WWW at http://:www.aboutwysetc.org/Docs/New_HorizonsII.pdf. Accessed 20.8.08.

Richards, G. and King, B. (2003) Youth travel and backpacking. *Travel and Tourism Analyst* 6, 1–23.

Richards, G. and Wilson, J. (2003) Today's youth travellers: Tomorrow's global nomads. New horizons in independent youth travel and student travel, a report for the International Student Travel Confederation (ISTC) and the Association of Tourism and Leisure Education (ATLAS). International Student Travel Confederation (ISTC), Amsterdam.

Richards, G. and Wilson, J. (2004a) Travel writers and writers who travel: Nomadic icons for the backpacker subculture? *Journal of Tourism and Cultural Change* 2 (1), 46–68.

Richards, G. and Wilson, J. (2004b) The international student travel market: Travelstyle, motivations and activities. *Tourism Review International* 8, 57–67.

Richards, G. and Wilson, J. (2004c) The global nomad: Motivations and behaviour of independent travellers worldwide. In G. Richards and J. Wilson (eds) *The Global Nomad: Backpacker Travel in Theory and Practice* (pp. 14–42). Clevedon: Channel View Publications.

Richards, G. and Wilson, J. (2004d) Widening perspectives in backpacker research. In G. Richards and J. Wilson (eds) *The Global Nomad: Backpacker Travel in Theory and Practice* (pp. 253–279). Clevedon: Channel View Publications.

Richards, G. and Wilson, J. (2006) Developing creativity in tourist experiences: A solution to the serial reproduction of culture? *Tourism Management* 27, 1209–1223.

Richardson, T. and Jensen, O.B. (2008) How mobility systems produce inequality: Making mobile subject types on the Bangkok sky train. *Built Environment* 34 (2), 218–231.

Riley, P. (1988) Road culture of international long-term budget travellers. *Annals of Tourism Research* 15 (3), 313–328.

Rivers, N., Lavin, M., Towne, D. and Zapaa, M. (2008) *Hostelling International – Boston: Economic and Social Impact Study*. Boston, MA: Centre for Public Management, Suffolk University.

Rodenburg, E. (1980) The effects of scale in economic development: Tourism in Bali. *Annals of Tourism Research* 7 (2), 177–196.

Rogerson, C.M. (2004) Regional tourism in South Africa: A case of 'mass tourism of the South'. *GeoJournal* 60, 229–237.

Rogerson, C.M. (2006) Pro-poor local economic development in South Africa: The role of pro-poor tourism. *Local Environment* 11, 37–60.

Rogerson, C.M. (2007a) The challenges of developing backpacker tourism in South Africa: An enterprise perspective. *Development Southern Africa* 24, 425–443.

Rogerson, C.M. (2007b) Backpacker tourism in South Africa: Challenges and strategic opportunities. *South African Geographical Journal* 89, 161–171.

Rogerson, C.M. (2007c) Reviewing Africa in the global tourism economy. *Development Southern Africa* 24, 361–379.

Rogerson, C.M. (2008a) Shared growth and tourism small firm development in South Africa. *Tourism Recreation Research* 33, 333–338.

Rogerson, C.M. (2008b) Backpacker tourism: Policy lessons for South Africa from the international experience. *Africa Insight* 37 (4), 27–46.

Rogerson, C.M. (2008c) Developing tourism SMMEs in South Africa: The need to recognize difference. *Acta Academica* 40 (4), 113–138.

Rogerson, C.M. and Kiambo, R. (2007) The growth and potential of regional tourism in the developing world: The South African experience. *Development Southern Africa* 24, 505–521.

Rogerson, C.M. and Visser, G. (eds) (2004) *Tourism and Development Issues in Contemporary South Africa*. Pretoria: Africa Institute of South Africa.

Rogerson, C.M. and Visser, G. (2006) International tourist flows and urban tourism in South Africa. *Urban Forum* 17, 199–213.

Römhild, R. (2002) Practised imagination. Tracing Transnational Networks in Crete and beyond. *Anthropological Journal on European Cultures* (11), 159–190.

Roth, T. (2001) *Community Marketing*. Gay and Lesbian Travel demographic report, 12 March.

Rough Guide (1999) *Sin Fronteras Australia*. London: Rough Guides (Spanish edition).

Ryan, C. and Hall, C.M. (2001) *Sex Tourism: Travels in Liminality*. London: Routledge.

Ryan, C. and Mohsin, A. (2001) Backpackers attitudes to the 'outback. *Journal of Travel and Tourism Marketing* 10 (1), 69–92.

Saffery, A. (2000) Mongolia's tourism development race: Case study from the Gobi Gurvansaikhan National Park. In P.M. Goode, M.F. Price and F.M. Zimmerman (eds) *Tourism and Development in Mountain Regions* (pp. 255–274). Wallingford: CABI.

Salazar, N.B. (2006) Touristifying Tanzania: Local guides, global discourse. *Annals of Tourism Research* 33 (3), 833–852.

SAT (South African Tourism) (2002) *Tourism Growth Strategy*. Johannesburg: SAT.

SAT (2004) *Global Competitiveness Project: Summary of Key Findings of Phase 1*. Johannesburg: SAT.

SAT (2007a) *Tourism in SA's Performance in the Past 5 Years*. Johannesburg: SAT.

SAT (2007b) *Gearing Up to be Globally Competitive: Tourism Growth Strategy 2008–2010*. Johannesburg: SAT.

SATOUR (South African Tourist Corporation) (1975) *Annual Report for the Year Ending 31st March 1974*. Pretoria: SATOUR.

Scheyvens, R. (2002a) *Tourism for Development: Empowering Communities*. Harlow: Prentice Hall.

Scheyvens, R. (2002b) Backpackers and local development in the Third World. *Annals of Tourism Research* 29 (1), 144–164.

Scheyvens, R. (2002b) Backpackers are beautiful; Assessing the potential of budget travellers to support community tourism in Africa. Unpublished proceedings of the Twentieth New Zealand Geography Conference, Palmerston North, July, pp. 278–282.

Scheyvens, R. (2006) Sun, sand and beach fale: Benefiting from backpackers – the Samoan way. *Tourism Recreation Research* 31 (3), 84.

Schivelbusch, W. (1986) *The Railway Journey: Trains and Travel in the Nineteenth Century*. Oxford: Blackwell.

Schofield, P. and Thompson, K. (2007) Visitor motivation, satisfaction and behavioral intention: The 2005 Naadam Festival, Ulaanbaatar. *International Journal of Tourism Research* 9 (5), 329–344.

Severin, T. (1991) *In Search of Genghis Khan*. New York: MacMillan.

Shaffer, T.S. (2004) Performing backpacking: Constructing 'authenticity' every step of the way. *Annals of Tourism Research* 24 (2), 139–160.

Sharpley, R. (2003) *Tourism, Tourists and Society*. Huntingdon: Elm Publications.

Shulman, S., Blatt, S.J. and Walsh, S. (2006) The extended journey and transition to adulthood: The case of Israeli backpackers. *Journal of Youth Studies* 9 (2), 231–246.

Shaw, G. and Williams, A. (2004) *Tourism and Tourism Spaces*. London: Sage.

Sheller, M. (2001) The mechanisms of mobility and liquidity: Re-thinking the movement in social movements. Department of Sociology, Lancaster University, Lancaster. On WWW at http://www.comp.lancs.ac.uk/sociology/papers/Sheller-Mechanisms-of-Mobility-and-Liquidity.pdf. Accessed 4.3.08.

Sheller, M. and Urry, J. (2003) Mobile transformations of 'public' and 'private' life. *Theory, Culture & Society* 20 (3), 107–125.

Sheller, M. and Urry, J. (2006) The new mobilities paradigm. *Environment and Planning A* 38 (2), 207–226.

Shipway, R. (2000) The international backpacker market in Britain: A market waiting to happen. In J. Swarbrooke (ed.) *Motivations, Behaviour and Tourist Types, Reflections on International Tourism* (pp. 393–416). Sunderland: Centre of Travel and Tourism.

Sibley, D. (2001) The binary city. *Urban Studies* 38, 239–250.

Simpson, K. (2004a) 'Doing development': The gap year, volunteer-tourists and a popular practice of development. *Journal of International Development* 16 (5), 681–692.

Simpson, K. (2004b) Broad horizons? Geographies and pedagogies of the gap year. PhD thesis. On WWW at http://www.gapyearresearch.org/recentresearchpublications.htm.

Singer, M. (2006) Investors wake up to sleepy fishing town. *The Sun-Herald*, 26 April, p. 66.

Slaughter, L. (2004) Profiling the international backpacker market in Australia. In G. Richards and J. Wilson (eds) *The Global Nomad: Backpacker Travel in Theory and Practice* (pp. 168–179). Clevedon: Channel View Publications.

Sönmez, S., Apostolopoulos, Y. and Tarlow, P. (1999) Tourism in crisis: Managing the effects of terrorism. *Journal of Travel Research* 38, 13–18.

Sørensen, A. (1992) Travellers i Sydøstafrika – en etnografisk introduction. Unpublished thesis. Århus University, Moesgaard.

Sørensen, A. (1999) *Travelers in the Periphery: Backpackers and Other Independent Multiple Destination Tourists in Peripheral Areas.* Nexo: Research Center of Bornholm.

Sørensen, A. (2003) Backpacker ethnography. *Annals of Tourism Research* 30, 847–867.

Speed, C. (2007) Are backpackers ethical tourists? In K. Hannam and I. Ateljevic (eds) *Backpacker Tourism: Concepts and Profiles* (pp. 54–81). Clevedon: Channel View Publications.

Star, S.L. (1999) The ethnography of infrastructure. *American Behavioral Scientist* 43, 377–391.

State of Queensland (2004) *Seventeen-seventy Locality – Zoning Plan.* May 2004.

Stewart, S. (2002) *In the Empire of Genghis Khan.* Guilford, CT: The Lyons Press.

Strüver, A. (2004) 'Space oddity': A thought experiment in European cross-border mobility. In J.O. Bærenholdt and K. Simonsen (eds) *Space Odysseys: Spatiality and Social Relations in the 21st Century* (pp. 63–82). Aldershot: Ashgate.

Sutcliffe, W. (1998) *Are You Experienced?* London: Penguin.

Swart, G. (2006) Flashpackers do it in style. *Sydney Morning Herald,* 22 February.

Sydney Morning Herald (2007) Brit ban on 'bloody' ad incredibly ludicrous. Travel section, 28 March.

Symes, C. (2007) Coaching and training: An ethnography of student commuting on Sydney's suburban trains. *Mobilities* 2 (3), 443–461.

Szerszynski, B. and Urry, J. (2002) Cultures of cosmopolitanism. *The Sociological Review* 50 (4), 461–481.

Teo, P. and Leong, S. (2006) A postcolonial analysis of backpacking. *Annals of Tourism Research* 33 (1), 109–131.

The Career Break Site (2008) On WWW at http://www.thecareerbreaksite.com/about-careerbreaks. Accessed 2.11.08.

The Economist (2006) Life beyond pay. Special report: Work-life balance. 17 June, p. 78.

The Future Laboratory (2004) *Spirit of Adventure: A Future Laboratory Insight Report for Standard Life Bank.* London: The Future Laboratory.

Thompson, C. and Tambyah, S. (1999) Trying to be cosmopolitan. *Journal of Consumer Research* 26, 214.

Thrift, N. (1996) *Spatial Formations.* London: Sage.

Thrift, N. (2000) Performing cultures in the new economy. *Annals of the Association of American Geographers* 90, 674–692.

Tomory, D. (1996) *A Season in Heaven: True Tales from the Road to Kathmandu.* London: Thorsons.

Tourism Australia (2006) Who is my market? Understanding the changing nature of backpacker. ATEC Symposium, 18 May.

Tourism Australia (2008) *Snapshot – Backpackers in Australia 2007.* On WWW at http://www.tra.australia.com/content/documents/Snapshots/2008/BackPacker_07_FINAL.pdf.

Tourism Research Australia (TRA) (2005) Backpackers in Australia, June 2005, *Niche Market Snapshots.* On WWW at www.tourismaustralia.com.au. Accessed 19.9.06.

Tourism Research Australia (TRA) (2008) *Snapshot: Backpackers in Australia 2007.* Canberra: Tourism Australia.

Travel Independent (2008) Backpacking travel advice – Southern Africa. On WWW at http://www.travelindependent.info/africa-southern.htm. Accessed 2.4.08.

Trauer, B. and Ryan, C. (2005) Destination image, romance and place experience – an application of intimacy theory in tourism. *Tourism Management* 26, 481–491.

Travelblogs.com (2009) 10 reasons to go flashpacking the next time you travel. On WWW at http://www.travelblogs.com/articles/10-reasons-to-go-flashpacking-the-next-time-you-travel. Accessed 19.7.09.

Travoholic (2008) *Flashpacking – What the Hell is it?* On WWW at http://Travoholic.com/articles/flashpacking.htm. Accessed 1.11.08.

Trembath, R. (2008) *Backpacker Travellers in South Australia: A Study of Itinerary Planning.* Southport, Queensland: Griffith University Sustainable Tourism Cooperative Research Centre.

Truco, T. (2004) Courting gay travellers. *The New York Times.* On WWW at http//:www.newyorktimes.com. Accessed 8.2.09.

Tucker, H. (2003) *Living with Tourism: Negotiating Identities in a Turkish village.* London: Routledge.

Tucker, H. (2007) Performing a young people's package tour of New Zealand: Negotiating appropriate performances of place. *Tourism Geographies* 9 (2), 139–159.

Turner, V. (1977) Variations on a theme of liminality. In S.F. Moore and B.G. Myerhoff (eds) *Secular Ritual* (pp. 36–52). Assen: Van Gorcum.

Uriely, N. and Reichel, A. (2000) Working tourists and their attitudes to hosts. *Annals of Tourism Research* 27 (2), 267–283.

Uriely, N., Yonay, Y. and Simchai, D. (2002) Backpacking experiences: A type and form analysis. *Annals of Tourism Research* 29, 519–537.

Urry, J. (1990) *The Tourist Gaze: Leisure and Travel in Contemporary Societies.* London: Sage.

Urry, J. (1995) *Consuming Places.* London: Routledge.

Urry, J. (2000) *Sociology Beyond Societies: Mobilities for the Twenty-First Century.* London: Routledge.

Urry, J. (2002a) Mobility and proximity. *Sociology* 36 (2), 255–274.

Urry, J. (2002b) *The Tourist Gaze* (2nd edn). London: Sage.

Urry, J. (2003) Social networks, travel and talk. *The British Journal of Sociology* 54 (2), 155–175.

Urry, J. (2005) The complexities of the global. *Theory, Culture and Society* 22 (5), 235–254.

Urry, J. (2006) Preface: Places and performances. In C. Minca and T. Oakes (eds) *Travels in Paradox: Remapping Tourism.* Lanham, MD: Rowman & Littlefield.

Urry, J. (2007) *Mobilities.* Cambridge: Polity Press.

Vance, P. (2004) Backpacker transport choice: A conceptual framework applied to New Zealand. In G. Richards and J. Wilson (eds) *The Global Nomad: Backpacker Travel in Theory and Practice* (pp. 237–252). Clevedon: Channel View Publications.

van Egmond, T. (2007) *Understanding Western Tourists in Developing Countries.* Wallingford: CABI.

Van der Duim, V.R. (2007) Tourism, materiality and space. In I. Ateljevic, N. Morgan and A. Pritchard (eds) *The Critical Turn in Tourism Studies: Innovative Research Methodologies* (pp. 149–163). Amsterdam: Elsevier.

Veal, A.J. (1993) The concept of lifestyle: A review. *Leisure Studies* 12, 233–252.

Veijola, S. and Valtonen, A. (2008) The body in the tourism industry. In A. Pritchard, N. Morgan, C. Harris and I. Ateljevic (eds) *Tourism and Gender: Embodiment, Sensuality and Experience* (pp. 13–31). Wallingford: CABI.

Virillo, P. (2005) *Negative Horizon*. London: Continuum.

Visser, G. (2003) The local development impacts of backpacker tourism: Evidence from the South African experience. *Urban Forum* 14, 264–293.

Visser, G. (2004) The developmental impacts of backpacker tourism in South Africa. *Geojournal* 60 (3), 283–299.

Visser, G. (2005) The local development impacts of backpacker tourism: Evidence from the South African experience, In E. Nel and C.M. Rogerson (eds) *Local Economic Development in the Developing World: The Experience of Southern Africa* (pp. 267–295). London: Transaction Press.

Visser, G. and Barker, C. (2004) Backpacker tourism in South Africa: Its role in an uneven tourism space economy. *Acta Academica* 36, 97–143.

Vogt, J. (1976) Wandering: Youth and travel behaviour. *Annals of Tourism Research* 4 (1), 25–40.

Waitt, G. and Markham, K. (2006) *Gay Tourism: Culture and Context*. New York: The Haworth Press.

Watts, L. (2007) The Art and craft of train travel. *Journal of Social and Cultural Geography* 9 (6), 711–726.

Watts, L. and Urry, J. (2008) Moving methods, travelling times. *Environment and Planning D: Society and Space* 26 (5), 860–874.

Waverley Council (2002) Visitor and Tourist Management Strategy. Sydney.

Wearing, B. (1998) *Leisure and Feminist Theory*. London: Sage.

Wearing, B. and Wearing, B. (1996) Refocusing the tourist experience: The flaneur and the chorister. *Leisure Studies* 15, 229–243.

Wearing, S. (2003) Special issue on volunteering. *Tourism Recreation Research* 28 (3), 1–104.

Wearing, S. (2004) Examining best practice in volunteer tourism. In R. Stebbins and M. Graham (eds) *Volunteering as Leisure/Leisure as Volunteering: An International Assessment* (pp. 209–224). Wallingford: CABI.

Weaver, D. (1995) Alternative tourism in Montserrat. *Tourism Management* 16 (8), 593–604.

Welk, P. (2004) The beaten track: Anti-tourism as an element of backpacker identity construction. In G. Richards and J. Wilson (eds) *The Global Nomad – Backpacker Travel in Theory and Practice* (pp. 77–91). Clevedon: Channel View Publications.

Welk, P. (2007) The Lonely Planet myth: 'Backpacker Bible' and 'Travel Survival Kit'. In K. Hannam and I. Ateljevic (eds) *Backpacker Tourism: Concepts and Profiles* (pp. 82–94). Clevedon: Channel View Publications.

Westerhausen, K. (2002) Beyond the beach: An ethnography of modern travellers in Asia. *Studies in Asian Tourism No. 2*. Bangkok: White Lotus.

Westerhausen, K. and Macbeth, J. (2003) Backpackers and empowered local communities. Natural allies in the struggle for sustainability and local control? *Tourism Geographies* 5 (1), 71–86.

Wheeler, T. (1986) *Australia – A Travel Survival Kit*. South Yarra: Lonely Planet.

White, L. (1949) *The Science of Culture: A Study of Man and Civilization*. New York: Farrar, Straus and Giroux.

White, T. (2008) Sex workers and tourism: A case study of Kovalam Beach, India. In J. Cochrane (ed.) *Asian Tourism: Growth and Change* (pp. 285–294). Oxford: Elsevier.

White, N.R. and White, P.B. (2004) Travel as transition: Identity and place. *Annals of Tourism Research* 31 (1), 200–218.

Wilde, H. (2006) Backpacker tourism in Global Sydney. Project Report, University of Western Sydney.

Williams, A.M. and Hall, C.M. (2002) Tourism, migration, circulation and mobility: The contingencies of time and place. In C.M. Hall and A.M. Williams (eds) *Tourism and Migration: New Relationships Between Production and Consumption* (pp. 1–52). Dordrecht: Kluwer.

Williams, A.M., King, R. and Warnes, A. (2004) British second homes in Southern Europe: Shifting nodes in the scapes and flows of migration and tourism. In M. Hall and D. Muller (eds) *Tourism, Mobility and Second Homes: Between Elite Landscape and Common Ground* (pp. 97–112). Clevedon: Channel View Publications.

Willis, K.S., Chorianopoulos, K., Struppek, M. and Roussos, G. (2007) Shared encounters. *Conference on Human Factors in Computing Systems archive. CHI'07 Extended Abstracts on Human Factors in Computing Systems* (pp. 2881–2884). San Jose, CA.

Wilson, J., Fisher, D. and Moore, K. (2007) 'Van Tour' and 'Doing a Contiki': Grand backpacker tours of Europe. In K. Hannam and I. Ateljevic (eds) *Backpacker Tourism: Concepts and Profiles* (pp. 9–25). Clevedon: Channel View Publications.

Wilson, E. and Little, D.E. (2005) A 'relative escape'? The impact of constraints on women who travel solo. *Tourism Review International* 9, 155–175.

Wilson, D. and Little, D.E. (2008) The solo female travel experience: Exploring the geography of women's fear. *Current Issues in Tourism* 11 (2), 167–186.

Wilson, J. and Richards, G. (2004) Backpacker icons: Influential literary 'Nomads' in the formation of backpacker identities. In G. Richards and J. Wilson (eds) *The Global Nomad: Backpacker Travel in Theory and Practice* (pp. 123–145). Clevedon: Channel View Publications.

Wilson, J. and Richards, G. (2008) Suspending reality: An exploration of enclaves and the backpacker experience. *Current Issues in Tourism* 11 (2), 187–202.

Wilson, J. and Richards, G. (2007) Suspending reality: An exploration of enclaves and backpacker experience. In K. Hannam and I. Ateljevic (eds) *Backpacker Tourism: Concepts and Profiles*. Clevedon: Channel View Publications.

Wilson, J., Richards, G. and MacDonnell, I. (2007) Intra-community tensions in backpacker enclaves: Sydney's Bondi Beach. In K. Hannam and I. Ateljevic (eds) *Backpacker Tourism: Concepts and Profiles* (pp. 199–214). Clevedon: Channel View Publications.

Worldtourism (2008) *Whitsundays History.* On WWW at http://worldtourism.com.au/Australia/QLD/Whitsundays/local-information/history.html.

Wrong, D.H. (1990) The influence of sociological ideas on American culture. In H.J. Gans (ed.) *Sociology in America* (pp. 19–30). Newbury Park: Sage.

Yeoman, I., Brass, D. and McMahon-Beattie, U. (2007) Current issues in tourism: The authentic tourist. *Tourism Management* 28, 1128–1138.

Yu, L. and Goulden, M. (2006) A comparative analysis of international tourists' satisfaction in Mongolia. *Tourism Management* 27, 1331–1342.